SpringerBriefs in Molecular Science

Green Chemistry for Sustainability

Series editor

Sanjay K. Sharma, Jaipur, India

More information about this series at http://www.springer.com/series/10045

Alice Meullemiestre · Cassandra Breil
Maryline Abert-Vian · Farid Chemat

Modern Techniques and Solvents for the Extraction of Microbial Oils

 Springer

Alice Meullemiestre
GREEN, UMR408, INRA
Avignon University
Avignon
France

Maryline Abert-Vian
GREEN, UMR408, INRA
Avignon University
Avignon
France

Cassandra Breil
GREEN, UMR408, INRA
Avignon University
Avignon
France

Farid Chemat
GREEN, UMR408, INRA
Avignon University
Avignon
France

ISSN 2191-5407   ISSN 2191-5415   (electronic)
SpringerBriefs in Molecular Science
ISSN 2212-9898
SpringerBriefs in Green Chemistry for Sustainability
ISBN 978-3-319-22716-0   ISBN 978-3-319-22717-7   (eBook)
DOI 10.1007/978-3-319-22717-7

Library of Congress Control Number: 2015947110

Springer Cham Heidelberg New York Dordrecht London

Printed on acid-free paper

Springer International Publishing AG Switzerland is part of Springer Science+Business Media
(www.springer.com)

Preface

Microorganisms, such as yeast, bacteria, and microalgae, are recognized as alternative sources of lipids due to their high productivity and rapid growth. Some microorganisms are able to accumulate more than 20 % of their dry cell mass in the form of lipids. In a particular and optimized culture conditions, some oleaginous yeast species can accumulate up to 70 % of their dry cell weight as lipids. Moreover, microbial oil exhibits a significant proportion of various fatty acids and thus offers a good potential as possible sources for biofuel or nutritional supplements from renewable resources. Scientific and industrial research laboratories are challenged to find an appropriate extraction technique, solvent, and procedure in order to disrupt cell's wall of microorganisms to free lipid's glands, with a minimum of energy consumption and promoting green solvent and innovative extraction techniques.

As a main difference to previously published books in this area, the readers such as chemists, biochemists, chemical engineers, physicians, and technologists will find a deep and complete perspective regarding extraction of microbial oil. The first part presents microorganisms as one of the most promising themes that can strongly contribute to green chemistry, not only in research laboratories, but also especially in various industries such as perfume, cosmetic, pharmaceutical, agricultural, food industries, and biofuel. The second part is dedicated to the importance of chemical analytical techniques and the need of standardization for quantitative and qualitative analysis of microbial oils. The last part gives new directions for research and industry by presenting conventional extraction techniques (Folch, Bligh and Dyer, Soxhlet) and intensified green extraction processes (ultrasound, microwave, supercritical fluid extraction, bio-based solvent, mechanical extraction, enzyme-assisted extraction, instant controlled pressure drop, and pulse electric field).

This work has been done as part of ProBio3 project financed by Agence National de la Recherche (France) to achieve scientific and technological challenges to boost the sustainable microbial production of lipids as biojet fuel and fine chemical compounds. Authors are totally convinced that this book is the starting point for future collaborations and innovations in a new area of "green chemistry" of "natural products" between research, industry, and education.

Acknowledgments

This study was funded by French Government Investissement d'Avenir ANR PROBIO3 (2012–2020). Cassandra BREIL thanks the FIDOP/FASO funds (Fonds d'Action Stratégique des Oléagineux) from the French vegetable oil and protein production industry for her doctoral grant.

Contents

About the Authors

Alice Meullemiestre is a postdoctoral researcher in the laboratory for green extraction techniques of natural product at the University of Avignon in France. Born in France (1986), she received her Ph.D. (2014) in process engineering from the University of La Rochelle and her "Maître de Conférences" qualification (2015) in process engineering and industrial chemistry. Her past and current researches are based on the intensification of various extraction processes (e.g., ultrasound, manothermosonication, microwave, subcritic, and instant controlled pressure drop) for green extraction of valuable compounds and lipids and transfer phenomena implied during S/L extraction. She has published eight papers in peer-reviewed journals and six communications for scientific meetings on this topic.

Cassandra Breil is a doctoral researcher of natural product chemistry from the GREEN Extraction Team, Avignon University in France. She received her master of green chemistry as specialization from University of Savoie. Her current research interest is the integration of innovative techniques (e.g., ultrasound, microwave, and subcritic) and agro-based solvents (e.g., vegetable oils and terpenes) for green extraction of lipids from microorganisms valued as food supplements or biofuel.

 **Maryline Abert-Vian** was born in 1974 and she received her Ph.D. (2000) in organic chemistry at the University of Avignon. She spent 4 years (2000–2004) as junior researcher with industrial companies. In 2005, she moved to the University of Avignon (France) to start her independent academic career. She obtained her "Habilitation à Diriger des Recherches" in 2011 in food and natural product chemistry, since she managed several French programs in the field of research and industrial application of alternative solvents applied for extraction of valuable compounds and biofuels from microorganisms (microalgae, yeast, etc.) with several industrial partners such as Airbus or GDF-Suez. Her research activity is documented by more than 25 scientific peer-reviewed papers and about 30 communications for scientific meetings, nine book chapters, and two patents. Her research primarily focuses on the solvent extraction and analysis of natural products and has paved the way for new extraction techniques with bio-based solvents.

 **Farid Chemat** is a full professor of chemistry at Avignon University, director of GREEN Extraction Team (alternative techniques and solvents), codirector of ORTESA LabCom research unit Naturex-UAPV, and scientific coordinator of "France Eco-Extraction" dealing with dissemination of research and education on green extraction technologies. His main research interests are focused on innovative and sustainable extraction techniques, protocols, and solvents (especially microwave, ultrasound, and bio-based solvents) for food, pharmaceutical, fine chemistry, biofuel, and cosmetic applications. His research activities have been documented by more than 150 scientific peer-reviewed papers, nine books, and seven patents.

Chapter 1
Microorganisms: Source of High Value Added Compounds

Abstract This chapter presents the potential of primary and secondary metabolites produced by microorganisms after a brief introduction of their classification and detailing their cell wall structure. Furthermore, applications of microorganisms as sources of reagents and ingredients in various industries: food, cosmetic, pharmaceutics, nutraceutics, and biofuel.

Keywords Microalgae · Yeasts · Bacteria · Fungi · Eukaryote · Prokaryote

Microorganisms have been present on earth since billions of years, and estimated to 2.5×10^{30} cells on Earth. The distribution of microorganisms on earth surface is heterogeneous: 66 % of marine subsurface, 26 % terrestrial subsurface, 4.8 % surface soil, 2.2 % oceans, and 1 % in all other habitats. Microorganisms could tolerate extremophile environment such as high temperature, high pressure, high salinity, and high radiation. They can evolve, survive and adapt themselves in new environments due to their ability to mutate and to transfer their genes. Some of them are pathogenic but many are inoffensive and mainly vital to human [1]. There are two classes or families of microorganisms: Eukaryotes and prokaryotes represented in Fig. 1.1.

We could find fungi and yeasts, microscopic plants as green algae, protists, animals among the eukaryotes and bacteria, archaea, virus in the family of prokaryotes. Only four on them are used to produce biomass: bacteria, microalgae, yeasts and fungi. This classification of microorganisms exists to obtain information concerning their shape, structure, displacement, way of eating, way of reproduction and propagation. Eukaryote is a multicellular or unicellular organism that contains several organelles: nucleus, mitochondrion, chloroplast, and Golgi apparatus. Eukaryote and prokaryote cells have 1 identical characteristics such as ADN, ARN and cell membrane. Cell wall structure and composition differ for each type of microorganisms as presented in Fig. 1.2.

Cell wall represent 50, 30 and 20 % respectively for microalgae, yeasts and bacteria. Microalgaes are composed mainly of neutral sugar (24–74 %) and with other minor compounds such as uranic acid (1–24 %), proteins (2–16 %), lipids (2–10 %) and glucosamine (15 %). Yeasts are composed mainly of glucan (29–33 %)

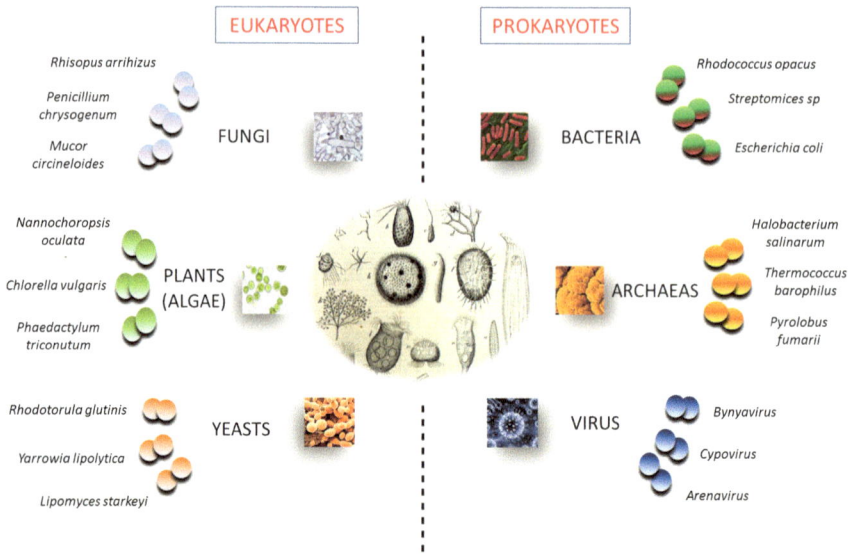

Fig. 1.1 Examples some eukaryotes and prokaryotes microorganisms

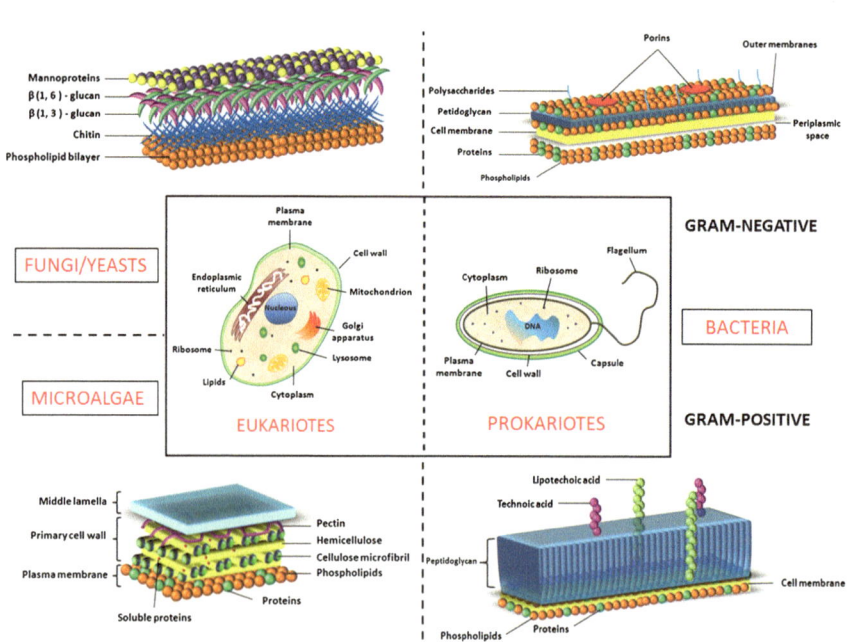

Fig. 1.2 Cell wall structure of some microorganisms

and mannan (31 %), followed by other minor compounds such as lipides (8–10 %), minerals (3 %) and chitin (3 %). The composition of cell wall differs from Gram-positive bacteria which are more resistant and composed of peptidoglucan (40–60 %), and from Gram-negative bacteria which are composed from lipids, proteins and peptidoglycan [2–5].

Microorganisms like microalgae, yeasts, bacteria and fungi have been used to produce primary and secondary metabolites for food, chemical, agriculture and pharmaceutical industries (Fig. 1.2). With the development of genetic sciences, many wild strains are genetically modified techniques have been used to increase quantity, quality and specified target molecules used in industry [7–8] (Fig. 1.3).

Microalgae possess a wide reserve of high value components of interest for cosmetic, food, pharmaceutics and biofuel applications such as colors, aromas, antioxidants, texturing agents... Some microalgae such as *Arthrospira* and *Chlorella* have been used in the skin care market (anti-aging, regenerate, anti-irritant cream and water binding agents). Others microalgae have been used for their organic metabolites such as sporopollein, scytonemin and amino acids which are protective agent against UV radiation. Microalgae is also a source of protein as food supplement used mainly in Asian countries as Japan or China. *Dunaliella salina* is used mainly for its high content in carotenoids such as β-carotene and zeaxanthin used as color, antioxidant or dietary supplement [9–11]. *Spirulina maxima* or *Haematococcus pluvialis* are cultivated for their minerals and vitamins beneficial for health. Others pigments like phycobiliproteins are extracted from *Porphyridium cruentum* and *Synechococcus spp.* [12]. Phycocyanin, a blue photosynthetic pigment, results from extraction of *Spirulina* [13] and sold as dietary

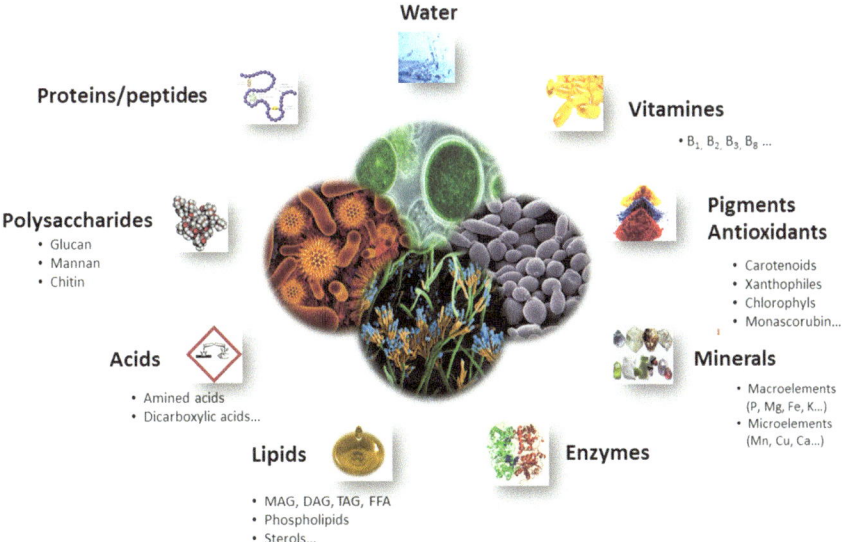

Fig. 1.3 Primary and secondary metabolites produced by microorganisms

supplement. Microalgae are also cultivated for the production of lipids like poly-unsaturated fatty acids (ω3 and ω6), their confer proprieties preventive against some diseases. Even if these fatty acids are used as food additive or pharmaceutical supplement [14], after transesterification, they are transformed in biofuel.

Yeasts and Fungi are commonly used for many years for fermentation process of food and beverage. They convert corn and other carbohydrates vegetable into ethanol to make beer, wine. Nowadays, these microorganisms are able to accumulate high value added target molecules. *Rhodotorula* is used to make an important amount of carotenoids (zeaxanthin, β-caroten, torulen) [7], lipids, proteins and organic acids. Fungi are largely used in pharmaceutical industry to produce various probiotic, antibiotic and antifungal properties by interfering with the human body [8].

Around 30 % of bacteria appear to be pathogenic and cause many diseases such as tuberculosis, typhus… The medical sector studies and employs bacteria to detect susceptible molecules to create vaccines or medications from their ability to remain at different germs. They are also used to manufacture antibiotics and others bioactive compounds such as insulin or antibiotics. On the nutrition point, they are able to metabolize vitamin such as vitamin B5 but also essential elements for the proper functioning of the nervous system, muscles and skin regeneration. In food industries, bacteria are employed for their capacities to produce lactic acid and are used to make yogurt, cheese, sour cream, buttermilk and other fermented milk products and vinegar. Bacteria produce wide grade of molecules qualified as pigments such as carotenoids, melanin, flavin and bacteriochlorophylls. Bacteria also involves in the environment domains such as treatments of wastewater, composting and landfills. Bacteria decompose themselves into compost or fertilizer and can be used in agriculture. During this process, methane gas is produced and valorized.

Figure 1.4 represents the different applications of microorganisms in different sectors such as food, nutraceutical, cosmetic, and pharmaceutics applications.

Microorganisms are considered as attractive and alternative source of lipids. Major lipid classes are present in cell wall and organelles of microorganisms. Microorganism's culture begins in Germany at the end of the First World War to produce oil [6]. During the Second World War, fermentation processes have evolved thus allowing having more information about a potential oil source other than plants. Since that time, Microorganisms have shown that they are suitable candidates for the production of natural products.

Oleaginous microorganism is an organism able to accumulate fat reserves exceeding 20 % of the dry weight of the cell, mainly triacylglycerol's. However, the phenomenon of accumulation of lipids is mainly seen in eukaryotic cells which can accumulate high quantities of lipids as some microalgae, filamentous fungi, bacteria and yeast [34]. The quantities of lipids differ according to microorganisms: strain, nature of the substrate, operating conditions and culture. It could be noticed that some yeast, fungi, microalgae and bacteria are able to produce a large quantity of oils rich of value added compound. These quantities can reach until 70 % of their biomass weight (Fig. 1.5) with certain cultivation conditions [35]. For these reasons

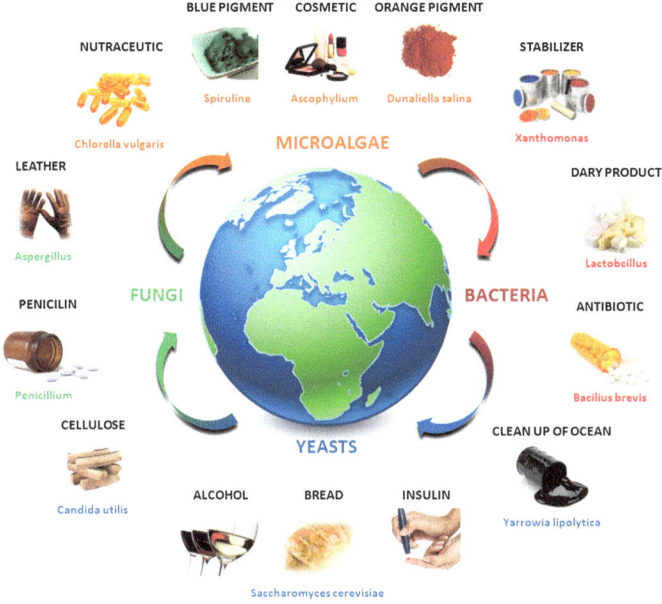

Fig. 1.4 Examples of applications from microorganisms

	MICROORGANISMS	STRAINS	MOLECULES OF INTEREST	APPLICATIONS	REFERENCES
COSMETIC	Bacteria	Chromobacterium	Violacein	Pigment	[9]
		Xanthomonas sp	Xanthane	Texture	[10]
	Microalgae	Arthrospira	Alginate, carraggeenen	Texture agent	[11]
		Nannochloropsis	Lutein, Neoxanthin, β-carotene	Pigment, antioxydant	[12]
		Tetraselmis	Fatty acids	Texture	[13]
		Chlorella	Sporopollenin	Protect themselves from UV	[14]
FOOD	Bacteria	Bacillus subtilis	Riboflavin	Vitamine in milk and energy drink	[9]
		Flavobacterium	β-carotene	Food additive	[15]
		Bradyrhizobium sp	Canthaxanthin	Colorant	[16]
		Corynebacterium glutamicum	Acid glutamic	Food additive	[1]
	Fungi Yeast	Aspergillus sp	Amylase	Baking agent	[17]
		Saccharomyces cerevisiae	Invertase / Pectinase	Candy / Wine, fruit juice	[18]
		Mucor pusillus	Rennin	Coagulation of milk	[19]
	Microalgae	Spirulina	Phycocyanin / Nutriments	Antioxidant	[7]
		Dunaliella salina	β-carotene	Antioxidant	[20]
		Porphyridium cruentum	Proteins	Food additive	[21]
PHARMACEUTIC NUTRACEUTIC	Bacteria	Duganella sp	Violacein	Antibiotic	[22]
		Serratia marcescens	Prodigiosin	Immunosuppressive, anticancer	[23]
		Bacillus sp	Bacitracin	Antibiotic	[24]
		Streptomyces griseus	Cycloserine	Antibiotic	[25]
		Propionibacterium	Vitamin B12	Supplement food	
	Fungi Yeast	Penicillium chrysogenum	Penicillin / Glucose oxydase	Drug / Antibiotic	[1]
		Aspergillus sp	Penicillin	Drug / Antibiotic	[1]
	Microalgae	Spirulina	Phycocyanin	Anticancer, supplement food	[7]
		Euglena gracilis	Biomin	Vitamins	[7]
ENVIRONMENT	Fungi Yeast	Rhodotorula glutinis	Fatty acids	Biofuel	[26]
		Yarrowia lipolytica	Fatty acids	Biofuel	[27]
		Yarrowia lipolytica	use metabolism	Cleanup, waste recycling, crop protection	[28]
	Microalgae	Nannochloropsis occulata	Fatty acids	Biofuel	[29]
		Anabaena	Proteins	Fertilizer	[7]

microbial microorganisms confers significant advantageous for the fourth generation of biofuel in comparison with the second and third biofuel generation (vegetable oils, animal fats):

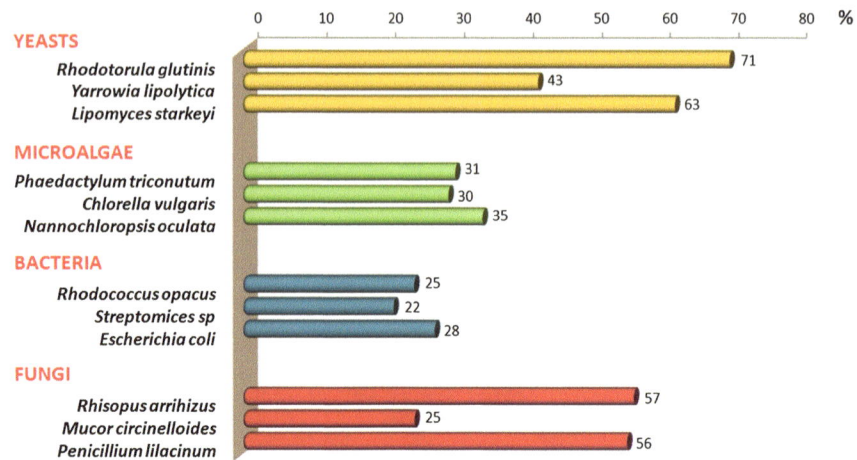

Fig. 1.5 Percentage of lipids from several microorganisms

- Ease of culture
- Rapid growth
- Reduction of contamination risk
- Culture are not dependent on weather
- Production at moderate temperature
- Development on renewable nutriments: food and agricultural waste
- Great metabolic diversity: can be cultivated to produce many molecules
- Extraction, separation and purification of simple of molecules.

Lipids in general and microbial oils in particular are classified into different categories based according to their polarities. Neutral lipids include acylglycerols: monoacylglycerol (MAG), diacylglycerol (DAG), triacylglycerol (TAG), Free Fatty Acids (FFA) and alkyl chain. They are mainly esters of fatty acids and glycerol. Polar lipids including phospholipids (PL) such as phosphatidylserin, phosphatidic acid, phosphatidylcholine, phosphatidylethanolamine and phosphatidylinositol. Those are diacylglycerols have a phosphate group which is bonded to a polar hydroxylated compound [17]. Another class includes sterols such as sitosterol, ergosterol, lanosterol, cholesterol and campesterol, which are cholesterol esters. They possess three hexagonal cycles and one alcohol secondary function. With deficiency conditions, TAG accumulates and forms a reserve element for microorganisms while PL and sterol are present in the constitution of the microorganisms (Fig. 1.6).

There are not less than 600 yeast and 60,000 fungi identified but only 25 of yeasts and 50 of fungi are known to produce intracellular lipid [36]. Most oleaginous yeasts can accumulate until 20–70 %. This way production has undeniable advantages compared to chemical industries vegetable oils. The lipids accumulated

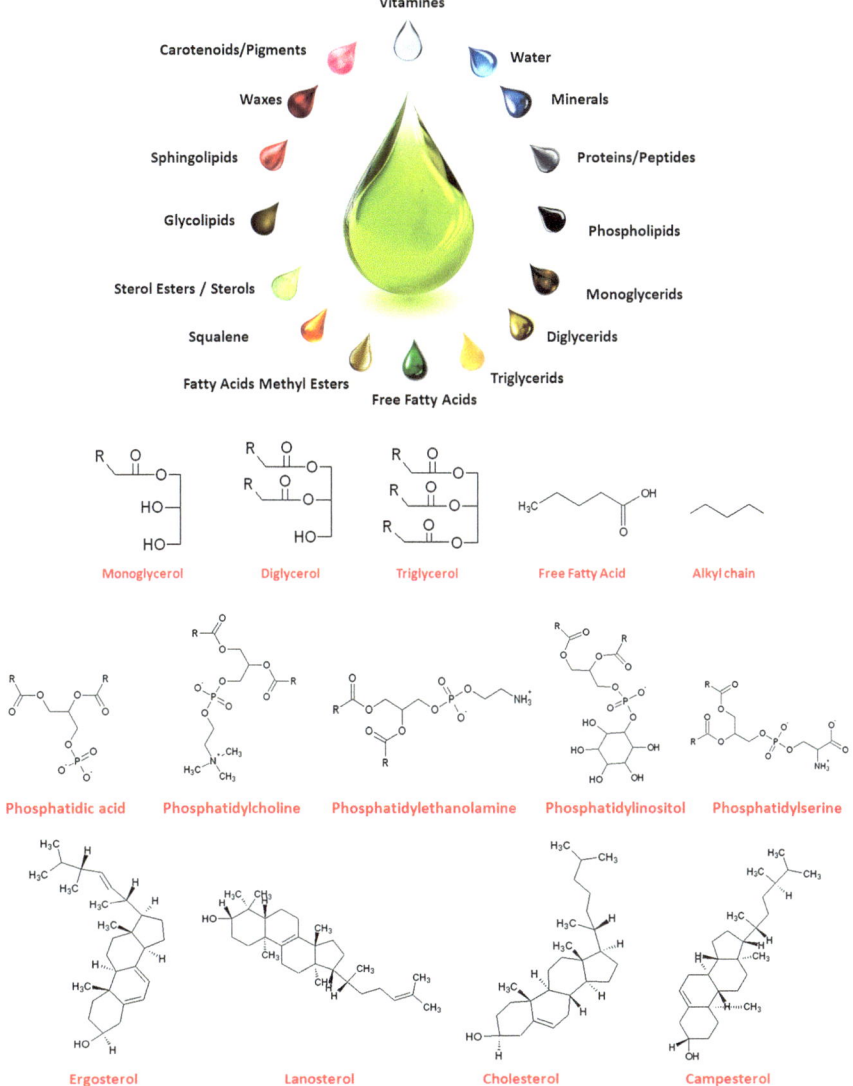

Fig. 1.6 Composition of microbial oils and chemical structures

inside these microorganisms and could be composed until 90 % of triglycerols and free fatty acids [37–41]. Sterols esters, sterols and phospholipids (important compounds of membrane structure) are minor components amounts. The fatty acids profiles are not the same even for the species, the choice of strain is very important for the domain concerned.

Microalgae can accumulate between 1 and 80 % of dry weight under particular conditions. Culture of microalgae requests some conditions of warm temperature,

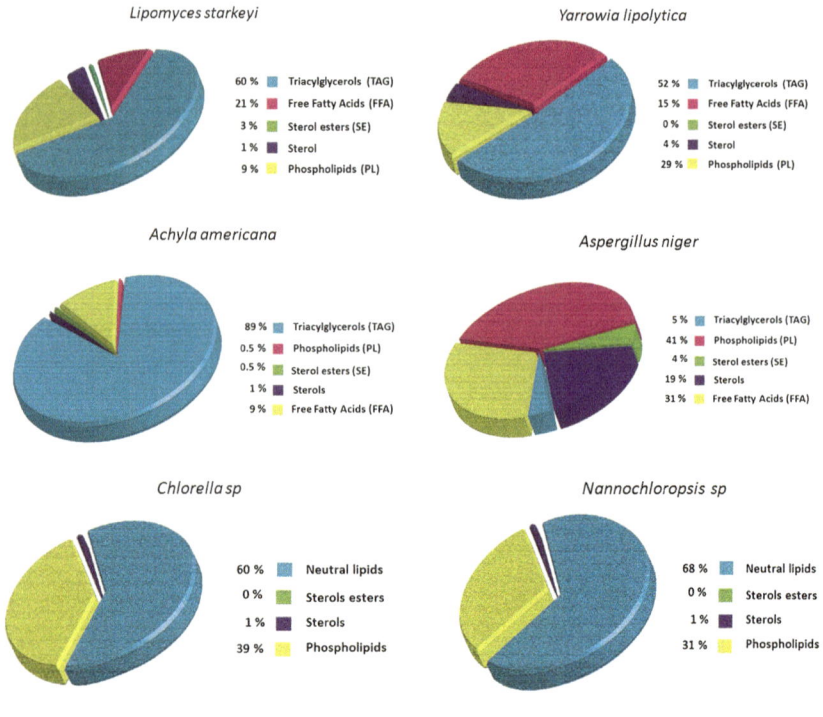

Fig. 1.7 Compositions of fatty acids and classes of lipids in microbial oils

The following data table is part of Fig. 1.7:

			Fatty acids composition (relative % w/w)						
	C14:0	C16:0	C16:1	C18:0	C18:1	C18:2	C18:3	others	References
Organism									
YEAST									
Rhodotorula glutinis	tr	17	1	2	42	21	9	9	[39]
Yarrowia lipolytica	tr	13	17	6	55	7	tr	2	[41]
Lipomyces starkeyi	tr	34	6	5	51	3	tr	1	[6]
MICROALGAE									
Phaedactylum tricornutum	9	27	45	1	5	tr	tr	C22:6 (13%)	[42]
Chlorella sp	tr	17	6	1	12	41	19	4	[43]
Nannochloropsis oculata	2	15	17	1	tr	1	1	C20:1 (57,5%)	[33]
BACTERIA									
Rhodococcus opacus	5	26	10	3	22	tr	tr	C15 (6%), C17 (13%), C17:1 (15%)	[44]
Streptomices sp	10	31	13	tr	tr	tr	tr	C15 (19%), others (27%)	[45]
Escherichia coli	4	37	9	4	5	tr	tr	C15 (12,2%), C17 (25,2%), C19 (2,9%)	[46]
FUNGI									
Rhisopus arrihizus	19	18	tr	6	22	10	12	13	[6]
Mucor circinelloides	tr	22	tr	5	38	10	15	10	[47]
Penicillium chrysogenum	3	13	tr	12	19	43	6	4	[37]

carbon dioxide, nutriments and sunlight to produce lipids. Their amounts depend of the quantities of nitrogen deficiency and others stress. As yeasts and fungi, the main lipids present in *Chlorella sp.* and *Nannochloropsis sp.* are triacylglycerols, free fatty acids and phospholipids. There are few bacteria oleaginous: *Nocardia sp.*,

Rhodococcus sp., *Mycobacterium sp.* and Arthrobacter which contain more than 50 % of lipids. Little information on the classes of lipids are identified so far [34]. Microorganisms are recognized for high productions of lipids. However, the chemical structure of fatty acids accumulated varies from a strain to another. Until recently, microorganisms were recognized a promising source for abundant amount of polyunsaturated fatty acids (PUFAs), which are used in medicines, food products or as biofuel. The most common carbon chains in microorganisms are polyunsatured fatty acid with 3 or 6 unsaturations with 16–18 carbons.

But some microorganisms have PUFAs with chain lengths ranging from 16 to 24 carbons with 2, 3, 4 untill 6 unsaturations such as 18:1, 18:2 and 18:3. Only one saturated fatty acids (C16:0) is dominating in yeasts, microalgae and fungi oil. The composition of fatty acids in microbial lipids from microorganisms could be determined by gas chromatography.

According to Fig. 1.7, short carbon chains like C8:0 to C10:0, are seen as trace. Medium and long chain of fatty acids as C12:0, C14:0, C20:0 and C22:0 are metabolize around 1 %. Bacteria have a atypical profile, fatty acids produced by bacteria are mainly of 16:0, 16:1 and 18:1 but also 15:0, 17:0 and 19:0, which are not found among microalgae, yeasts and fungi. Regarding the quantities of fatty acids, there is a real variation in the lipid profile according to microorganisms. These quantities of fatty acids depend on the culture conditions and/or modified genes so the choice of microorganism is crucial for intended application. Industrial applications of microbial oil depend on the fatty acid profile, on the carbon chain length and the degree of unsaturation of the chain. For example, unsaturated fatty acids such as arachidonic acid, γ-linolenic acid and linoleic acid have many important functions in the human body. Nowadays, these acids can be used in everyday life and are essential in cosmetic, medicine and food supplements. They are classified as high value fatty acid according to their market and efficient value. On the other hand, short and saturated carbon chains are more suitable to produce biofuel. Cultures of palm, rapeseed oil and soybean increase in size because of the production of biodiesel and fatty acids high added values. These last years, the cost of some product (pharmaceutical, food…) has increased. These circumstances have necessity to find new unconventional, non-edible and sustainable sources of fatty acids from microbial oils.

Chapter 2
Analytical Methodology for Lipid Extraction and Quantification from Oleaginous Microorganisms

Abstract This chapter reviews the development of extraction and analytical techniques for lipids from microorganisms. The commonly used extraction techniques of total lipid such as «Folch», «Bligh and Dyer», and Soxhlet Extraction methods are detailed and explained. A special focus has been made for the complementary analytical methods such as HP-TLC (High Performance Thin Layer Chromatography) which gives details about the repartition of the different classes of lipids and GC (Gas Chromatography) which brings quantitative analysis of all fatty acids chains present in the lipid microbial extract.

Keywords Folch method · Bligh and dyer method · Soxhlet extraction · High performance thin layer chromatography · Gas chromatography

Analytical tools for assessment of lipids from microorganisms have become important in comparative and screening studies. Conventional protocols for analysis of lipids typically involve organic solvent extraction followed by gravimetric determination for total lipid content and/or chromatographic analysis for obtaining total fatty acids profile or lipids class determination. An accurate analysis of lipid content requires the complete extraction of lipids from biomass, which depends of the efficiency, and specificity of the procedure in use. Furthermore, lipid extraction methods for microorganisms' cells are not well established, and there is currently no real standard extraction method for the determination of the lipid content, classes and composition. For example, the absence of standard procedures for extraction and chemical analysis leads to controversial results in the area of microalgae biofuel research papers. In this part, we present approved standard methodologies to evaluate the total lipid content, total fatty acids profile and lipid classes for oleaginous microorganism sample, which can be in dry or wet form.

Analysis of lipids from oleaginous microorganisms are divided into several stages, according to Fig. 2.1, beginning with the extraction of total lipids which can be performed according to three different procedures such as Soxhlet [48] applied with dry samples, Bligh and Dyer [49], and Folch [50], with wet samples, following by (1) gravimetric lipid measurement, (2) transformation of lipids in methyl esters (derivatization) and, identification and quantification of such methyl esters

© The Author(s) 2015

A. Meullemiestre et al., *Modern Techniques and Solvents for the Extraction of Microbial Oils*, SpringerBriefs in Green Chemistry for Sustainability, DOI 10.1007/978-3-319-22717-7_2

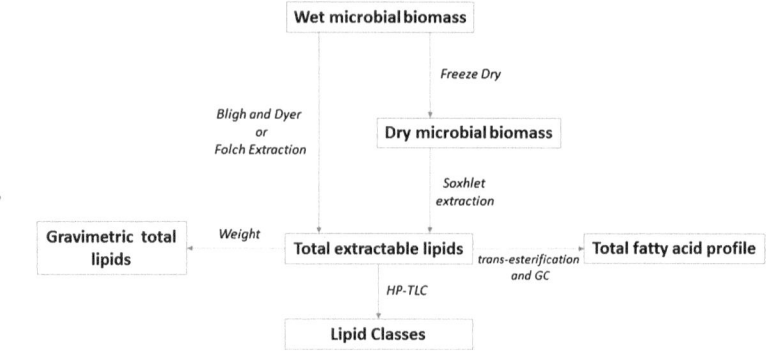

Fig. 2.1 Flow diagram of analysis procedure

(FAMEs, fatty acid methyl esters) by gas chromatography, (3) analysis of different lipid classes performed by High Performance Thin Layer Chromatography.

2.1 Commonly Used Methods for Determination of Total Lipid Content

Various types of lipids occur in microorganisms. Simple lipids are often found in aggregates in storage tissues, from which they are relatively easily extractable. Complex lipids are usually constituents of biological membranes, where they are found associated with such compounds as proteins and carbohydrates and they are extracted laboriously. These lipids are linked with other components by weak interactions as Van der Waal's forces, Hydrogen bonds or ionic bonds [51]. The solubility of lipids in various solvents depends of the relative affinity between the solvent and either the hydrophobic or the hydrophilic parts of the molecule. Lipids such as triacylglycerides or sterol esters with functional group of low polarity are very soluble in hydrocarbon solvents like hexane or higher polar solvents like chloroform and ethers. They tend to be rather insoluble in polar solvent such as methanol. In contrast, polar lipids like phospholipids tend to be less soluble in hydrocarbon solvents, even although their dissolution can be helped by the presence of other lipids, but they dissolve easily in more polar solvents such as chloroform and methanol. So, to extract lipids from biological tissues, extraction solvent has to dissolve lipids readily but also disrupt weak interactions between lipids and tissue matrix.

During lipid extraction, microorganism biomass is exposed to an eluting extraction solvent or mixture of solvents, which extract the lipids out of the cellular matrices. Once the crude lipids are separated from the cell debris, extraction solvent and water (only when extraction is performed on wet biomass) could be evaporated to measure gravimetrically extract's yield. Lipid extractability varies for different microorganism species. Thus, there is no universal procedure however the most common used methods for determining total lipid content in micro-organisms are the Soxhlet [48, 52] extraction with hexane for dry samples and those using a mixture of chloroform and methanol described by Folch et al. [50] and Bligh and Dyer [49]. These methods remain the gold standards in academic and industrial research laboratories.

2.1.1   "Folch" Method

Folch et al. [20] were the first which develop the chloroform/methanol/water phase system (the so-called "Folch" method) for extraction of lipids from biological material. The method is still considered as the classical and most reliable method for quantitative extraction of lipids [53]. The method relies on a mixture of chloroform and methanol, forming a monophasic solvent system, to extract and dissolve lipids. A biphasic system is then produced by the addition of water or saline solution leading to the migration of polar compounds along with the methanol into an upper water phase and leaving the lipids in the lower chloroform phase. The endogenous water in the microorganism is associated with the upper aqueous phase.

The method uses a ratio of 1 part of sample to 20 parts of solvent with a ratio of 2:1 chloroform/methanol mixture followed by several washings of the crude extract with weak salt solutions of $NaCl/KCl/MgCl_2$ in order to retain acidic lipids.

2.1.2   Bligh and Dyer Method

Bligh and Dyer [49] is the most cited reference method in the literature for extraction of lipids from biological materials [54]. The Bligh and Dyer method is a simple adaptation of the Folch procedure and was developed simply as an economical way (in terms of solvent volumes) of extracting lipids from tissues such as fish muscle. The main differences between the protocols of Folch et al. and Bligh and Dyer are not only the volume of solvent system in relation to the amount of sample, but also the ratios between solvents within the systems, and the presence or absence of salt in the added water phase. While Folch et al. employed 20 times sample volume with a 2:1 (v/v) chloroform-methanol mixture, Bligh and Dyer used a chloroform-methanol extraction stage of 1:2 and 1:1 (v/v) amounting to a final volume of only four times the equivalent sample amount. According to Fig. 2.2, briefly, the wet microorganism biomass was placed to a glass tube with Teflon-lined screw cap and, after addition of 3 mL methanol-chloroform (2:1, v/v), was vortexed for 10 min. Then, 1 mL of chloroform and 0.8 mL of KCl 0.88 % (w/v) were added before vortexing and centrifuging at 4000 rpm for 5 min. The upper phase was discarded and the lower chloroform phase was transferred to a new glass tube. After evaporation, the lipid extract will be re-suspended in a small volume of solvent for HPTLC or GC analysis.

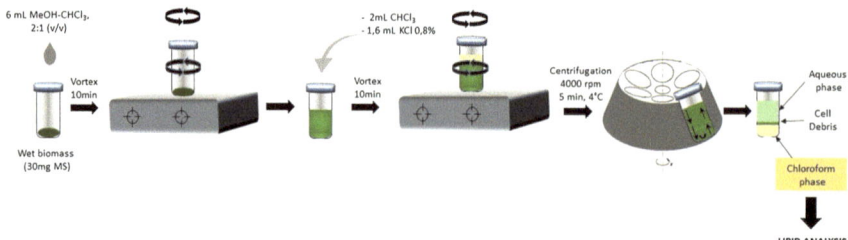

Fig. 2.2 Bligh and Dyer procedure

2.1.3   Soxhlet Extraction

The process and procedure described by Soxhlet [55] in 1879 is the most popular method used for lipid extraction from food and natural products. According to Soxhlet's procedure, lipids are extracted from solid material by repeated washing or leaching (percolation) with a fresh organic solvent, usually hexane or petroleum ether, under reflux in a special glassware. Hexane is also one of the most desirable hydrocarbon solvent designed for triacylglycerol extraction [56]. In this method the sample is dried, ground into small particles and placed in a porous cellulose thimble. According to Fig. 2.3, the thimble is placed in an extraction chamber (2), which is suspended above a distillation flask containing the solvent (1) and below a condenser (4). The flask is heated and the solvent evaporates and moves up into the condenser where it condensed into liquid that trickles into the extraction chamber

Fig. 2.3 Soxhlet apparatus

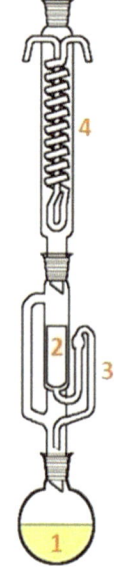

containing the sample. When the solvent surrounding the sample exceeds an upper level of the extractor siphon it overflows and trickles back down into the distillation flask. At the end of the extraction process, which usually takes several hours, the flask containing solvent and lipid is removed. In some device a funnel (3) allows to recover the solvent at the end of the extraction after closing a stopcock between the funnel and the extraction chamber. Solvent in the flask (1) is then evaporated and the mass of the remaining lipid is measured by gravimetry to calculate the amount of lipid in the initial sample.

2.2 Analysis of Lipids

The lipids that are available from microorganisms varied, and a detailed knowledge of the accurate composition is required for the development of potential future industrial applications. The following is an outline of some key methods used in the analysis of microbial lipids as well as the type of information that is obtained by these techniques.

Gravimetric measurement of the total lipids extracted from biomass is the simplest method of lipid characterization. Besides, this does not provide enough information about the constitution of microbial lipids. High Performance Thin Layer Chromatography (HP-TLC) and Gas Chromatography (GC) are complementary methods which provide more accomplished results. HP-TLC gives details about the repartition of the different classes of lipids and GC brings quantitative analysis of all fatty acids chains present in the lipid microbial extract.

2.2.1 Gas Chromatography Analysis

Gas chromatography is a sensitive analytical technique that can identify and quantitate transesterified fatty acids from saponified lipid species (Triglycerides, Phospholipids, Cholesteryl Esters and Free Fatty Acids). Lipid extracts must be volatilized prior to GC separation. Thus, the polar Fatty Acids are transformed into their less polar methyl esters derivatives (FAMEs) by methanolysis. Numerous methods and reagents exist for the transesterification of Fatty Acids, with the reagents generally being classified as either base- or acid-catalysed reaction. The most common techniques involve the strong acid BF_3 in methanol that is heated with the sample for up to 1 h [57]. Other acid reagents include H_2SO_4 [58–60] and HCl [61] in methanol. Base-catalysed reactions using $NaOCH_3$, KOH or NaOH are also popular, but are not efficient catalyzing reagents for non-esterified fatty acids [62]. Capillary columns and a Flame Ionization Detector (FID) are most commonly used to separate and detection FAMEs in biological materials. Identification of fatty acids is based on the retention times of FAMEs and the most common fatty acids are available in a commercial reference mixture. To confirm the identity of each component, Gas Chromatography-Mass Spectrometry (GC-MS) can be used in order to compare the masse spectrum with a reference spectrum according to commercial and approved databases.

2.2.2 High Performance Thin Layer Chromatography (HP-TLC)

Lipids from microorganisms can be separated into the various classes by HP-TLC. It is an analytical technique based on Thin Layer Chromatography separation with major improvements such as better resolution and accurate quantitative analysis of compounds. HP-TLC has been reported to provide excellent separation, qualitative and quantitative of numerous biomolecules from natural sources or biological samples [63, 64]. HPTLC plates have much finer particles than regular TLC plates, and consequently they are better suited for quantitative analysis. Common lipid separation and quantitation techniques are generally time consuming, cost intensive, concerning the amounts of solvents, and are not suitable for routine analysis of several samples [65]. HP-TLC is an offline process in which various stages are carried out independently. A typical procedure consists of three main steps: sample application, chromatogram development and chromatogram evaluation as illustrated in Fig. 2.4. Sample application is an important and critical step for obtaining good resolution for quantification of compounds. For quantitative TLC, automated sample application is necessary. Typically, samples are loaded on silica gel HP-TLC plates through bands with the spray on technique using the Automatic TLC Sampler 4 (Camag, Switzerland).

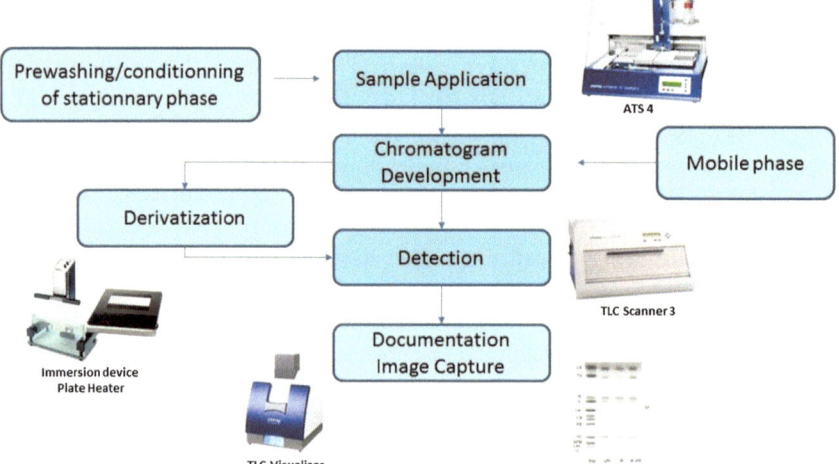

Fig. 2.4 HP-TLC complete system and procedure

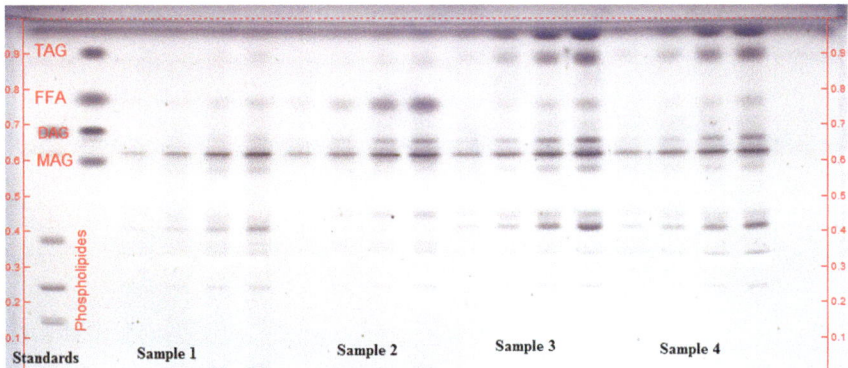

Fig. 2.5 HPTLC chromatogram, where the different lipid classes detected can be visualised and their order of appearance in microalgae extracts (four tracks per samples), *MAG* monoacylglycerols, *DAG* diacylglycerols, *FFA* free fatty acids, *TAG* triacylglycerols

After application stage, the chromatogram is developed by dipping in suitable solvent system taken in automatic developing chamber, ADC 2 (Camag, Switzerland), which is the heart of an HP-TLC system. It allows the development step fully automatically, reproducibly, and independent of environmental effects. In microbial lipid analysis, HP-TLC involves developing neutral and polar lipids in one dimension on the same chromatogram using different solvent systems in succession [60]. The plate is first developed in solvent system I consisting of "methyl acetate:isopropanol:chloroform:methanol:KCl" (0.25 % solution) (25:25:25:10:9, by volume) running to a height of 5.5 cm from the origin. After drying, the plate is developed in solvent system II (n-hexane/diethyl ether/glacial acetic acid mixture (80:20:2) to a height of 8.5 cm from the origin. After drying, the plate is dipped in a modified $CuSO_4$ reagent (20 g $CuSO_4$, 200 mL methanol, 8 mL H_2SO_4, and 8 mL H_3PO_4) then heated at 141 °C for 30 min on a TLC plate heater [60]. After development and derivatization, lipids are detected by charring and quantified using a CAMAG 3 TLC scanning densitometer (CAMAG, Switzerland) with identification of the classes against known polar and neutral lipid standards. The densitometry date are reported as values which are expressed as percent of lipid class in microbial lipids. TLC Vizualiser (CAMAG, Switzerland) allows to capture images of HP-TLC plate as can be seen on Fig. 2.5.

Chapter 3
Innovative Techniques and Alternative Solvents for Extraction of Microbial Oils

Abstract This chapter reviews the recent development of extraction techniques, procedures and solvents for lipids from microorganisms. The modern innovative and intensified extraction techniques, alternative solvents and original procedures (ultrasound, microwave, supercritical fluid extraction, biobased-solvent, mechanical extraction, enzyme-assisted extraction, Instant controlled pressure drop, pulse electric field) are summarized in terms of their principles, processes, applications, benefits and disadvantages.

Keywords Intensification · Extraction · Innovative techniques · Original procedures · Alternative solvents

3.1 Ultrasound-Assisted Extraction

3.1.1 Principle

Ultrasound is mechanical vibrations characterized by their frequency range (from 20 kHz to 10 MHz), and can be differentiated into two types depending on their frequency: diagnostic and power ultrasound.

- Diagnostic ultrasound (low power and high frequency), which frequencies are comprised between 1 and 10 MHz and ultrasonic intensity below 1 W cm^{-2}. Its applications are enormous in medical field as diagnostic or control tools.
- Power ultrasound (high power and low frequency), with frequencies ranged from 20 kHz to 1 MHz, and ultrasonic intensity above 1 W cm^{-2} are used to produce physical or chemical effects into the medium. Its main applications are sonochemistry (in order to facilitate or accelerate chemical reactions), agriculture (water dispersion) or in industry (cutting, plastic welding).

Power ultrasounds are able to generate cavitation bubbles. The chemical or physical effects caused by ultrasound in the extraction medium are attributed to the

© The Author(s) 2015
A. Meullemiestre et al., *Modern Techniques and Solvents for the Extraction of Microbial Oils*, SpringerBriefs in Green Chemistry for Sustainability,
DOI 10.1007/978-3-319-22717-7_3

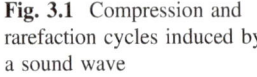

Fig. 3.1 Compression and rarefaction cycles induced by a sound wave

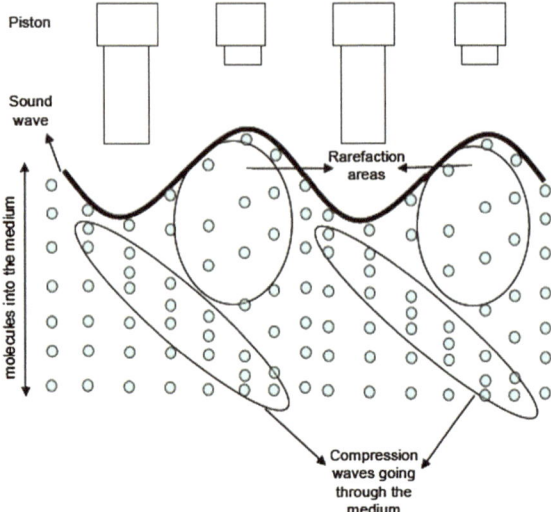

cavitation phenomena. As the sound wave spreads through an elastic medium, it induces a longitudinal displacement of particles, acting as a piston on the medium surface (Fig. 3.1), resulting in a succession of compression and rarefaction phases [66].

The ultrasound wave induces cavitation bubbles, which are able to grow during rarefaction cycles, and decrease in size during compression phases [67]. When the bubbles reach a critical diameter, they collapse during a compression cycle inducing a release of large amounts of energy. The temperature and pressure reached during bubble collapsing have been estimated to be up to 5000 K and 20 MPa (inside a simple ultrasonic bath at room temperature). The hot spots generated are able to accelerate significantly the chemical reactivity into a medium. If the cavitation bubbles are formed near to a solid surface, the implosion resulting, which induces release of high pressure and temperature, generates micro-jets and shock waves directed towards the solid surface [68]. In the case of biological compounds (Fig. 3.2), the cavitation bubble generated close to the material surface collapses during a compression cycle and a micro-jet directed toward the surface is created.

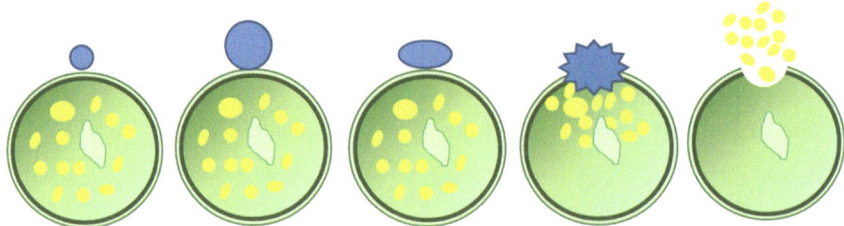

Fig. 3.2 Collapse of cavitation bubble

The high pressure and temperature involved in this process will destroy the cell walls of the plant matrix and its content will be released into the medium.

3.1.2   Process and Procedure

For batch sonication, two classical types of equipment are used for extraction purposes: ultrasonic baths and ultrasound probes (Fig. 3.3). Ultrasonic bath has been the first equipment used in sonochemistry, the frequencies generated are ranged from 25 to 50 kHz and ultrasonic intensity comprised between 1 and 5 W cm^{-2}. The main advantages of ultrasonic bath is its price (inexpensive) and to have many applications such as sample preparation, cleaning, homogenization, solid dispersion and degassing. It can also be used for extraction but the low power of ultrasound delivered directly to the matrix to be treated and the low reproducibility are the major disadvantages [69]. Ultrasonic probe is directly immersed into

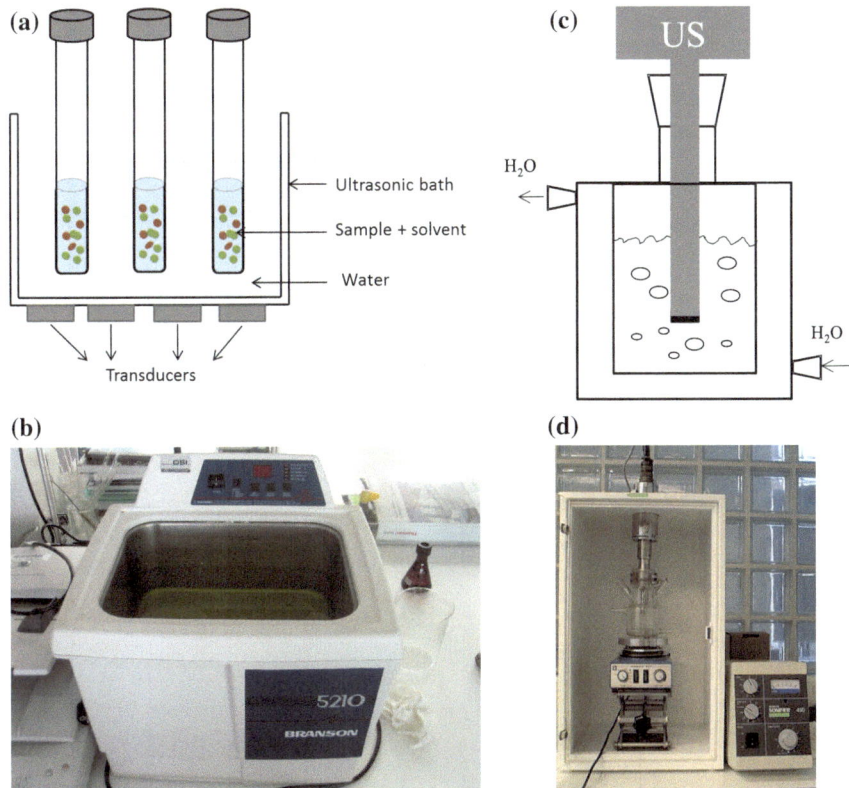

Fig. 3.3 Laboratory ultrasound apparatus: ultrasonic bath (**a, b**) and ultrasonic probe (**c, d**)

the reactor, which enables a direct irradiation of medium, while generating acoustic pressure largely higher [69].

However, the use of ultrasound probes is limited to small volume of fluid and the system leads to a quick increase of temperature in the reactor. In the case of an extraction, then it is necessary to cool the reactor by a double-jacket, in order to keep a constant temperature inside the reactor.

For industrial applications, one of the major parameter is the quantity of product to be treated; the ultrasonic probe can handle only small volumes, so one of the solutions is to use continuous systems that can handle a larger amount with a restrictive volume of reactor, ultrasound are then more concentrated with a maximum power. The other alternative is to use ultrasonic baths with a larger radiating surface and an agitation system. REUS Company has developed a wide range of US bath from laboratory scale (0.5–3 L), to pilot (30–50 L) and industrial scale (500–1000 L) (Fig. 3.4). Ultrasounds can be coupled or combined with other

Fig. 3.4 Industrial ultrasound apparatus: **a** 3 L, **b** 30 L, **c** 300 L

extraction techniques such as microwave energy [70], supercritical fluid extraction [71] or with conventional methods such as Soxhlet extraction [72].

3.1.3   Application for Microbial Oil

Diversity of devices and depending on the desired ultrasound effect (mechanical or chemical) allow many applications of ultrasound technologies in industry. Ultrasound technology can be used to activate phase transfer phenomenon and are therefore used for homogenization, emulsification, crystallization, filtration and extraction [73]. Ultrasound is used to extract bioactive components such as essential oils, antioxidants, fat and oils, and colors from natural products, but also for biofuel extraction [74, 75]. More particularly, ultrasound-assisted extraction (UAE) has been extensively employed to extract lipids from oleaginous microorganisms such as bacteria [76, 77] and microalgae [78, 79] for biodiesel conversion. In fact, shock waves from cavitation bubbles created during ultrasound process allow disrupting of the cells and especially walls of microorganism, facilitating the diffusivity of solvent into the matrix and enhancing liberation of valuable components [80, 81]. Furthermore, a study conducted by Zhang et al. [76] shown that ultrasound allow reduction of extraction time from 12 h to 15 min in comparison with conventional solvent extraction without modification of fatty acids composition. Adam et al. [74] confirmed these results with ultrasound-assisted extraction of microalgae in a reduced time and same quality of fatty acids composition compared to conventional extraction methods, with a possible scalable industrial apparatus. In Table 3.1 are

Table 3.1 Ultrasound-assisted extraction of lipids from microorganisms

Matrix	Types	Extract	Experimental conditions	References
Trichosporon oleaginosus	Bacteria	Lipid	US horn, 520 kHz, 40 W, 15 min, 25 °C, Folch method MeOH/CHCl$_3$ (2:1, v/v)	[76]
			US bath, 50 Hz, 2800 W, 15 min, 25 °C, Folch method, MeOH/CHCl$_3$ (2:1, v/v)	
Nostoc sp.	Bacteria	Lipid	US bath, 50 Hz, 15 min, 30 °C, Bligh and Dyer, MeOH/CHCl$_3$ (2:1, v/v)	[77]
Tolypothrix sp.	Bacteria	Lipid	US bath, 50 Hz, 15 min, 30 °C, Bligh and Dyer, MeOH/CHCl$_3$ (2:1, v/v)	[77]
Chlorella spp.	Microalgae	Lipid	US Bath, 40 kHz, 29.7 W/L, Bligh and Dyer	[78]
Nannochloropsis	Microalgae	Lipid	US probe, 100 W, 50–60 °C, Bligh and Dyer, Folch method, MeOH	[79]

shown some examples of ultrasound-assisted extraction of lipid from oleaginous microorganisms for biodiesel production.

3.2 Microwave-Assisted Extraction

3.2.1 Principle

Ganzler and Lane [82, 83], in 1986, were the first to report microwave-assisted extraction technique. This technology was first used in the extraction of various compounds from food products (citrus, aromatic plants, cereal etc.). The process consisted to heat with microwave the vegetable matrix immerged in a highly absorbing solvent (such as methanol) for extraction of polar compound or in a low absorbing solvent (such as n-hexane) for extraction of non-polar compounds [84]. The mixture (matrix + solvent) was heated for short periods, without reaching boiling point, followed by cooling steps. When oil glands of the plant are subjected to severe thermal stress and localized high pressure, as in the case of microwave heating, the pressure building-up within the glands exceeds its capacity for expansion, causing their rupture more rapidly than for a conventional extraction (Fig. 3.5) [84]. Liberated metabolites are dissolved in the organic solvent and are further separated by liquid-liquid extraction.

Microwave extraction technique has been tested and approved for its effectiveness for extraction of several classes of compounds such as oils, fats, organic compounds, essential oils, antioxidants in terms of yield [85, 86] but most important factor is reduced extraction time resulting in a reduction of energy consumed. The process was patented by Paré et al. [87] from Environment Canada in 1990 as microwave-assisted process (MAP).

3.2.2 Process and Procedure

Microwave-assisted solvent extraction can be performed using two technologies: open-vessel system operating at atmospheric pressure and closed-vessel system under controlled temperature and pressure. Laboratory microwave focused

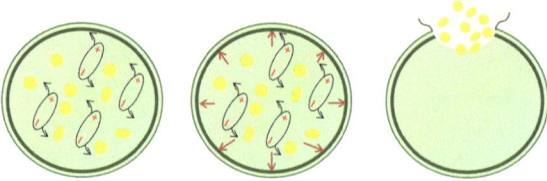

Fig. 3.5 Mechanism of microwave-assisted extraction

open-vessel system was introduced by Prolabo in 1986. It operated under atmospheric pressure; in consequence the maximal temperature that can be reached is determined by the boiling point of the solvent at atmospheric pressure and is represented in Fig. 3.6. The experiments with open-vessel can be realized either with a multi-mode, where microwave irradiation is allowed to disperse randomly in a cavity and the material is directly irradiated or with a single-mode where microwave irradiation is focused on a restricted zone; as a consequence the sample is subjected to a stronger electrical field compared to multi-mode systems [88]. The microwave focused energy combined to the advantages of Soxhlet extraction was developed by Luque de Castro et al. [89] as focused microwave-assisted solvent extraction (FMASE). Its principle is the same as a conventional Soxhlet extraction, the only point that sample cartridge is placed in the irradiation zone of focused microwave oven. The solvent is heated by microwave energy which ensures a homogeneous and a very efficient heating and heat diffusion through the sample. FMASE is appropriated for control quality analysis of fats and oils in plant or food products with a higher efficiency than conventional Soxhlet extraction in a shorted extraction time and a reduction of solvent and energy used.

Following the success of this technology, Chemat team proposed a modified microwave-integrated Soxhlet extraction (MIS). The sample is not in contact with the fresh solvent [90]. Usually, closed-vessel systems are based on multi-mode microwave and the sample is submitted to drastic conditions such as high extraction temperature. The solvent can be heated above its boiling point at atmospheric

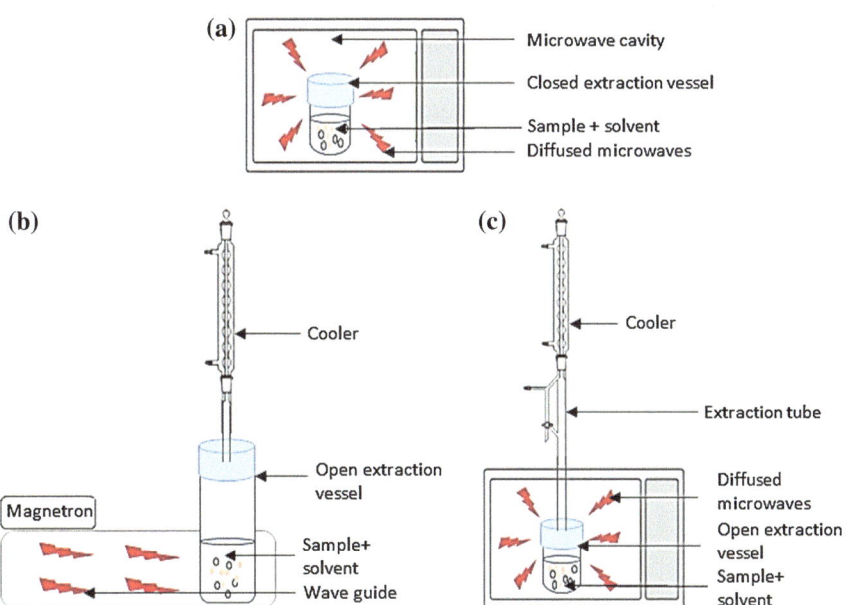

Fig. 3.6 Experimental set-up for laboratory microwave-assisted extraction apparatus

pressure; as a consequence the mass transfer of target compounds from the matrix to the solvent is accelerated due to the combined effect of pressure and high temperature [91].

More recently, a new procedure called 'simultaneous distillation and extraction process' (SDEP/MW) has been developed by Dejoye et al. [85], especially for lipid extraction from wet microalgae. This extraction method consists of four steps: elimination of water contained into the matrix, lipid extraction and elimination of solvent in a single "in situ" step. The main advantage of this new procedure is time saving compared to Soxhlet extraction which results in an energy gain and positif environmental impact.

3.2.3 Application for Microbial Oil

Lipid extraction of oleaginous microorganisms by microwave technology has been applied to various matrices, as shown in Table 3.2. Microorganisms have a rigid cell wall (mainly yeast strain), compared to other biological samples; in consequence it is more difficult to obtain a complete extraction process. Generally, conventional lipid extraction from microorganisms involves a preliminary step as cell disruption (to break or open cell wall) to improve extraction efficiency, that necessities two steps of preparation which generates a great time and energy consuming. Microwave-assisted extraction permits to combine the two unit operations in only step: cell disruption and extraction. Additionally microwave technology provide the same extraction efficiency as compared to conventional extraction but with an important reduction of extraction time for oleaginous yeast *Rhodothorula glutinis* lipid extraction from 4 h (Soxhlet) to 30 s using microwaves [92]; and for *Saccharomyces ceevisae* yeast lipid extraction from 360 min (Soxhlet) to 10 min using microwaves [93].

Table 3.2 Microwave-assisted extraction of lipids from microorganisms

Matrix	Types	Extract	Experimental conditions	References
Rhodotorula glutinis	Yeast	Lipid	MASE, 1 atm., 20 min	[92]
			MeOH/CHCl$_3$ (2:1, v/v)	
Saccharomyces cerevisae	Yeast	Lipid	MASE, 1 atm., 800 W, 60 and 80 °C, 10 min, MeOH/CHCl$_3$ (2:1, v/v)	[93]
Tolypothrix sp.	Bacteria	Lipid	MASE, 1 atm.	[94]
Nannochloropsis oculata	Microalgae	Lipid	MW/SDEP, 1 atm., 400 W, 10 min, Bligh and Dyer	[85]
Dunaliella salina				
Nannochloropsis gaditana	Microalgae	Lipid	MASE, 1 atm., 60 °C, 10 min	[95]
			MASE under pressure, 90 °C, 10 min	

3.3   Instant Controlled Pressure Drop

3.3.1   *Principle*

The instant controlled pressure drop technology, abbreviated DIC according to the French expression 'Détente Instantanée Contrôlée' developed by Allaf et al. in 1988 [96, 97]. The process is based on fundamental studies concerning the thermodynamics of instantaneity and auto-vaporization processes combining with hydro-thermo-mechanical evolution of many biopolymers for food, cosmetic, and pharmaceutical purposes [97]. DIC consists in a high temperature/high pressure-short time (HTST) treatment followed by an instant pressure drop toward vacuum, as shown in Fig. 3.7.

This fast transition from high-pressure steam to vacuum induces significant pressure drop in the vicinity of the matrix (Fig. 3.8). The low relative pressure of

Fig. 3.7 Pressure-time profile of DIC processing

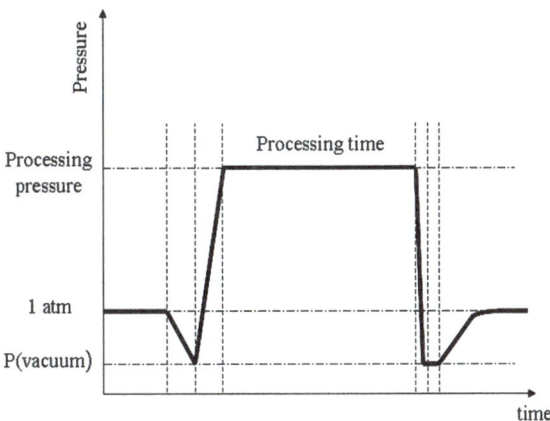

Fig. 3.8 Mechanism of DIC process

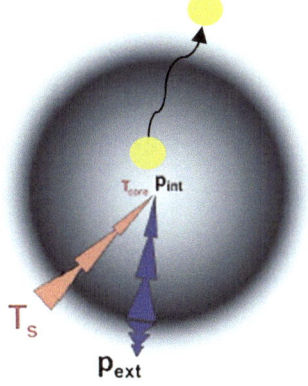

water vapor around the vacuum pressure involves an auto-vaporization. Its principle consists to submit the plant material to a steam pressure (P < 1 MPa) at a high temperature (180 °C) during a short time (order of few seconds) followed by an abrupt pressure drop to vacuum (3–5 kPa, Δt = 20–200 ms) [98]. This abrupt pressure drop ($\Delta P/\Delta t$ > 25 × 10^6 Pa^{-1}), which characterizes DIC technology, induces several effects. Firstly, it causes an auto-vaporization of water and a releasing of the complex liquid contained in the sample, and then it induces an instantaneous cooling of the sample. As well, the rapid pressure drop induces a significant mechanical stress related to auto-vaporization and swelling phenomenon, causing the rupture of cell and secretion walls but also the expansion of material [99]. The purpose of these effects leads to texture change which results in higher porosity as well as increased specific surface area and reduced diffusion resistance of the sample [100]. Experimental conditions of DIC extraction allows reduced processing time and the instant reducing temperature drop prevents further thermal deterioration and ensure a high quality of extract.

3.3.2 Process and Procedure

DIC equipment is composed of four major components, i.e., (1) an extraction vessel where the sample to be treated is placed; (2) a controlled pressure-drop valve, which ensures a quick and controlled liberated of steam pressure contained in the extraction vessel to the vacuum pump; (3) a vacuum system composed of a vacuum pump and tank with a volume 50-fold higher than the volume of the treatment vessel; (4) an extract collection trap used to recover condensates (Fig. 3.9).

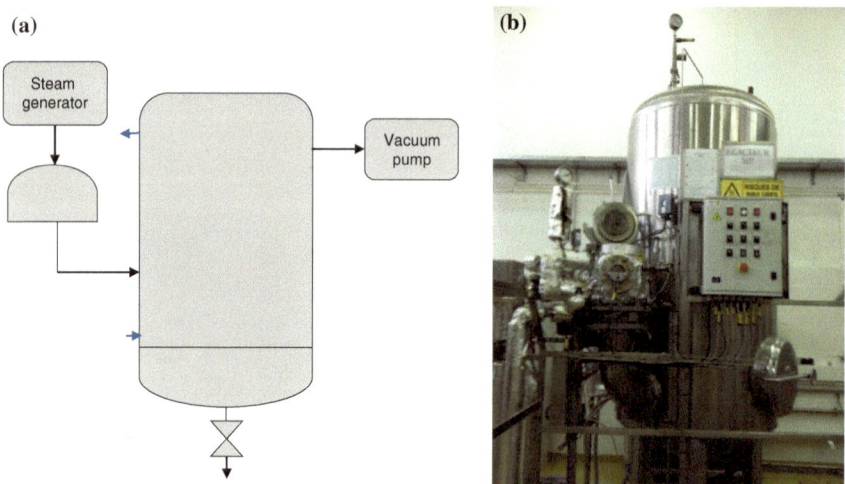

Fig. 3.9 Schematic representation (**a**) and photography (**b**) of DIC equipment

Classical procedure for extraction by DIC can be described as following. A certain quantity of humidified sample is placed in the treatment vessel at atmospheric pressure. Then an initial vacuum (0.5 MPa) is established to facilitate the penetration of the heating fluid in order to enhance heat transfer into the product. Then, the vessel is filled by an injection of saturated steam until the desired pressure (0.1–0.6 MPa) during a fixed processing time. After this period, controlled pressure-drop valve is instantaneously opened (in less than 0.2 s), resulting in an abrupt pressure drop inside the treatment vessel. After steam release, the atmospheric pressure is returned back inside the reactor and the mixture of condensed water and metabolites can be recovered from the vacuum tank.

3.3.3 Applications for Microbial Oil

Nowadays, DIC treatment is considerate as an efficiency method for lipid extraction from oleaginous material such as oilseed. Indeed, Nguyen [101] has proved in a study that DIC allow enhancing lipid extraction from jatropha and rapeseed seeds without significant modification of fatty acids composition in comparison with conventional Soxhlet extraction. Allaf et al. [102] have shown that enhancing of lipid extraction by DIC treatment is clearly noticed by calculation of effective diffusivity. In fact, effective diffusivity was higher after a DIC treatment of rapeseed seeds. More recently, DIC was endorsed as a pretreatment for in-situ transesterification in the case of microalgae: Phaeodactylum. Optimized DIC treatment (P = 0.16 MPa and t = 68 s) allows increasing of 27 % in total lipid and more than 75 % in FAMEs yield [103]. Additionally to lipid extraction, it was observed that the residual microalgae allows increasing of lutein extraction. Moreover DIC allows reducing the energy consumption and manufacturing cost compared to conventional processes of lipid extraction.

3.4 Supercritical Fluid Extraction

3.4.1 Principle

Fluid is considered as supercritical when its temperature and pressure are above its critical point; where distinction between liquid and gas phases does not exist [104]. Supercritical state is known for almost two centuries by a French scientist Baron Charles Cagniard de la Tour in 1822 [105]. In 1870, Andrews introduced the term 'critical' and estimated the particulars of critical point of carbon dioxide [106]. Hannay and Hogarth in 1879 and 1880 are demonstrated the capacity of supercritical fluids to solvate solids [107]. Supercritical fluid extraction (SFE) was emerged at the industrial stage in 1970 as an alternative process to plant distillation.

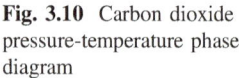 **Fig. 3.10** Carbon dioxide pressure-temperature phase diagram

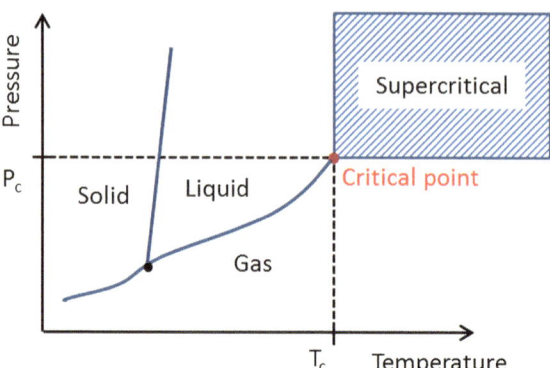

Extraction by a supercritical fluid is similar to a conventional solid-fluid extraction. The particularity of supercritical fluids is based on alteration of their physical properties, which can be modulated by an increase of pressure and/or temperature above their critical values. At these conditions the properties of some fluids can be placed between those of a gas and those of a liquid (Fig. 3.10).

Whereas the supercritical fluids owns a density similar to liquids, which induces apparition of solvating power, and a viscosity similar to gases, its diffusivity is intermediate between these two states, thus favoring the mass transfer between the solute to extract and the supercritical fluid. These conditions offer to supercritical fluids the solvent properties particularly interesting in the case of extraction. The supercritical carbon dioxide ($SC-CO_2$) is nowadays the mostly used and constituted more than 90 % of industrial applications, due to its low critical coordinates ($T_c = 31$ °C and $P_c = 7.38$ MPa) easily achievable in practice.

Moreover, it has significant advantages as to be inert, non-flammable, inexpensive, easily available, environment-friendly, and GRAS (Generally Recognized As Safe) solvent [108]. CO_2 is volatilized at atmospheric pressure induces that after depressurization extracts are free of solvents [109], this characteristic permits to remove one operation step of evaporation after extraction process which is the most energy consuming in industry.

3.4.2   Process and Procedure

SFE equipment's is generally composed of four main components, as shown in Fig. 3.11: (1) a volumetric pump, to ensure the pumping of fluid, which pump can be preceded by a cooler that brings gaseous component in a liquid state; (2) a heat exchanger; (3) an extractor, where pressure is established and maintained by a back pressure regulating valve; and (4) a separator, its number depends on the application, up to three can be placed in series to achieve multiple fractioning of the desired molecules contained in the extracts.

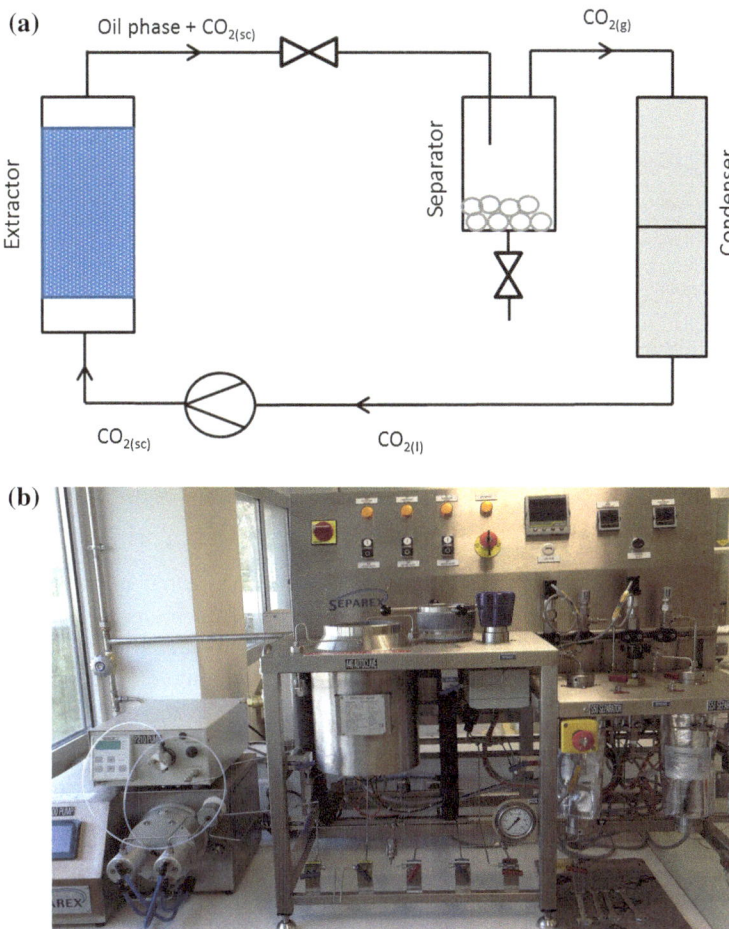

Fig. 3.11 Schematic representation (**a**) and photography (**b**) of supercritical fluid extraction installation

Extraction by supercritical fluids is composed of two major steps: an extraction step followed by a separation step of the solute from the solvent. First, the fluid is generally sequentially pressurized and heated to reach its supercritical state. The resulting supercritical fluid feeds the extractor with an ascending or descending flux. Then, supercritical fluid extracts the desired solute contained in the matrix. Inside the separator, supercritical fluid will return to its gaseous state, and the solute will be no longer solubilized and will be then separated by gravity. Finally, the extracts are collected at the bottom of the separator. Depending on the equipment, the gas can be recycled or re-injected in the system, or released in the atmosphere.

Table 3.3 SC-CO_2 extraction of lipids from microorganisms

Matrix	Types	Extract	Experimental conditions	References
Yarrowia lipolytica	Yeast	Lipid	40 °C, 20 MPa, 120 min compared to EtOH maceration during 20 h at T_{amb}	[110]
Saccharomyces cerevisiae	Yeast	Lipid	EtOH as co-solvent, 40 °C, 20 MPa, 120–180 min	[111]
Schizochytrium limacinum	Microalgae	Lipid	EtOH as co-solvent, 40 °C, 35 MPa, 30 min	[112]
Nannochloropsis sp.	Microalgae	Lipid + carotenoids	40 °C, 30 MPa, $Q_{CO_2} = 0.62$ g/min	[113]
Scenedesmus sp.	Microalgae	Lipid	50 °C, 50 Mbar, $Q_{CO_2} = 3$ ml/min	[114]

3.4.3 Applications for Microbial Oil

Supercritical CO_2 has been successful studied by many researchers for extraction of lipids and especially polyunsaturated fatty acids from microalgae [112–115], but few references [110, 111] has been found in literature about extraction of lipids by SC-CO_2 from yeasts. Nobre et al. [113] have indicated a more effective extraction using combined solvent SFE-CO_2/ethanol, allowing to extract 45 g of lipids per 100 g of dry microbial biomass and recover 70 % of pigments initially present (Table 3.3). These result have been confirmed by Milanesio et al. [110] that proved the efficiency of extraction by supercritical CO_2 for lipids from Y. lipolytica, recovering 10 mg of yeast oil per g of solvent in pure CO_2. It was noticed that addition of ethanol increase the concentration of yeast oil about 15 mg of yeast oil per g of solvent. These results have been confirmed by Nobre et al. [113] that indicate more ·effective extraction using combined solvent SFE-CO_2/ethanol, allowing to extract 45 g of lipids per 100 g of dry microbial biomass and recover 70 % of pigments initially present (Table 3.3).

3.5 Pulse Electric Field Extraction

3.5.1 Principle

Pulse electric field (PEF) has been used in genetic engineering for the incorporation of large DNA molecules and electro-fusion [116]. Since the discovery of the destructive effect on microorganisms by Doevenspeck in 1961 [117], PEF has become an attracting technique in food processing: preservation, transformation and extraction. Sale and Hamilton [118] were the first, in 1967, to study the bactericidal effect induced by pulse electric fields.

The principle of PEF technology is the application of short pulses of high electric fields during a short time (around micro to milliseconds). The processing time is calculated by multiplying the number of pulses times with effective pulse duration. The process is based on the potential formation of pores inside the cell membranes, due to their exposure to low-moderate external electric fields of adequate strength (<10 kV/cm) during a very short time. This phenomenon is called "electroporation" or "electropermeabilization" [119, 120]. Lebovka et al. [121] hypothesized that PEF treatment affects the cellular tissues as a correlated percolation through two processes: releasing of cells and moisture mass transfer inside the cellular structure, which is sensitive to PEF treatment repetitions. Thus electroporation mechanism increases the permeability of plant cells, which permit enhancement of solute diffusion through the cell membrane [122], as represented in Fig. 3.12. Major advantages of PEF extraction is short treatment time which allows energy consumption saving.

3.5.2   Process and Procedure

The application of pulsed electric field consists in treating a biological product, in batch or continuously, between two electrodes. Sample is submitted to a pulsed electric voltage, generally ranged between 15 and 80 kV/cm with very short time of pulses of 1–5 µs for microorganisms destruction; and 0.1–5 kV/cm with short pulses time of 5–1000 µs for electroporation of plant cells and non-thermic extraction from solid food. In most cases to submit a biological sample to an electric field, PEF system required four basic elements. First, a high voltage generator that supplies the electric power to the desired voltage. Next, several capacitors, which temporarily store electrical energy and a switch that allows the discharge of the stored energy. Then, the fourth and most fundamental element of PEF system is the treatment chamber, composed of two electrodes where the high voltage pulses are

Fig. 3.12 Mechanism of PEF treatment on biological cells

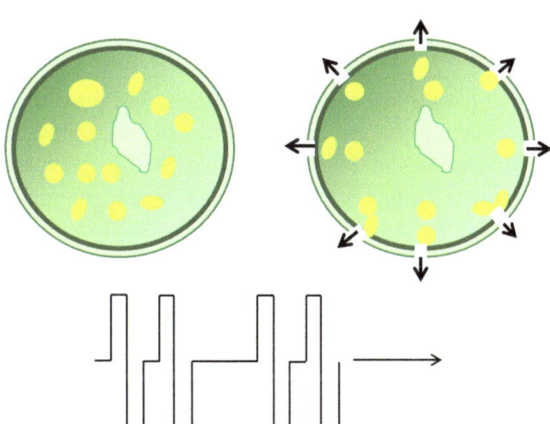

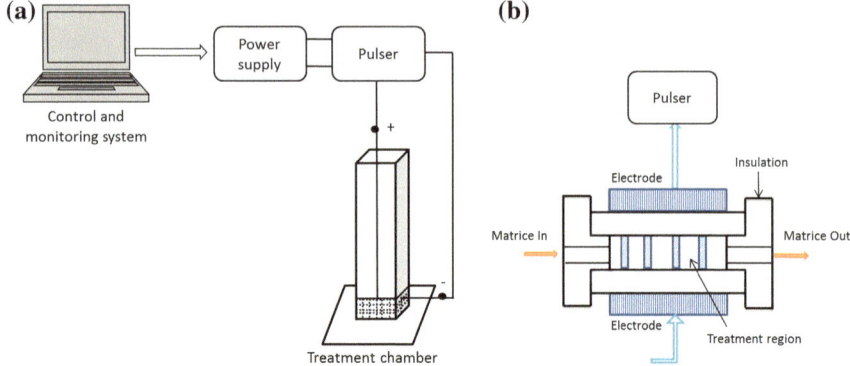

Fig. 3.13 Schematic representation of simplified PEF installation (**a**) and treatment chamber (**b**)

applied into the sample to be processed (Fig. 3.13b). There are two major types of treatment chamber: static chamber where only a given volume can be processed at a same time (for experimental application) and dynamic chamber which allows a continuously treatment, according to industrial requirement. On closing of switch, condensers discharge in the processing chamber giving rise to an electric pulse through the sample. The waveforms of pulses generated by PEF generators can be exponential decay, oscillatory, square, triangular, or more complex wave forms [123]. The exponential decay, triangular, and square pulses may be either mono-polar or bi-polar. Square-wave pulses have the advantage of applying the electric field to the desired tension during almost the entire pulse duration and to reduce the temperature rise [124]. Square-wave generators are more expensive and required more complex equipment than exponential decay generator. The alternating pulses offer minimum energy consumption with reduced the electrolytic deposition on the electrodes and smaller food electrolysis [60].

3.5.3   *Applications for Microbial Oil*

During the last decade, PEF-treatment was proved to be useful for intensifying the conventional processes for extraction of lipids from marine microorganisms. According to the Foltz study [126], PEF-treatment was proved as an efficient method to extract lipids from microalgae biomass such as *Chlorella vulgaris*, *Chlamydomonas reinhardtii* and *Dunaliella salina*. It was shown that a single 4 kV/cm electric field pulse was sufficient to lyse *C. vulgaris* cells. Author also concluded that additional PEF could be necessary to release lipids through the cell membrane and cellulose wall. At the same time, Zbinden et al. [127] shown the potential of PEF as a pre-treatment of cell disruption of *Ankistrodesmus falcatus* and enhance lipid recovery of 130 % compared to control sample, with reduced extraction time. PEF-treatment of biomass allows enhancing the mass transfer and

Table 3.4 Pulse electric field for extraction of lipids from microorganisms

Matrix	Types	Experimental conditions	References
C. vulgaris	Microalgae	2.7 kV/cm	[126]
		21 pulses 100 μs at 10 Hz	
Dunialiella sp.	Microalgae	4 kV/cm	[126]
		60 pulses	
A. falcatus	Microalgae	45 kV/cm	[127]
		360 ns 1/e pulse duration	
A. protothecoides	Microalgae	23–43 kV/cm	[128]
Synechocystis PCC 6803	Cyanobacteria	>35 kWh/cm^3	[129]

solvent diffusion due to the cell disruption. Another study conducted by Sheng et al. [129] has shown evidence of severe cell disruption of cyanobacteria such as *Synechocystis PCC 6803* after a PEF-treatment (>35 kWh/m^3) which enhanced the solvent transfer and favor the accessibility of lipids. This phenomenon was explained by the increasing of temperature during pulse electric field which causes destruction of auto-fluorescence compounds, as well as destruction of plasma, cell wall and thylakoid membranes. Thus, the application of pulsed electric field (PEF) may be a promising alternative to conventional extraction of lipids from microorganisms, for reducing extraction time, enhancing yields, lowering costs, and having positive environmental impacts (Table 3.4).

3.6 Mechanical Pressing Extraction

3.6.1 Principle

Pressing is a simple and ancestral method used for extracting oil from oleaginous seeds. Expression or pressing can be defined as the expelling of a fluid contained in a porous material by squeezing or compression forces [130]. Sometimes, in order to enhance yield of extraction, pressing is generally preceded by a pretreatment such as drying for example for algal biomass [131], which is energy consuming. In the case of expeller press or screw press, applied pressure in a particular range improves the oil extraction efficiency but too much pressure will generate a loss of lipids and heat generation [132]. The first expeller presses was invented and patented by Anderson in 1900 and was successful used in oil mills [132]. During hydraulic pressing, oil is expelled from material by a uniaxial force. The pressure applied is generally very high and this type of press is limited to a certain type of raw materials (cocoa, olives) [133]. Therefore in oil mills, hydraulic presses have been progressively replaced by screw pressing, allowing a continuous expression. During expeller pressing, a helical screw, in a barrel toward a restriction, conveys the raw material. In addition to a mechanical pressure on material, shear forces are developed along the screw allow expression of oil contained in material. This type of presses allows obtaining higher oil's yield. The compaction of the cake and the increasing pressure may have several sources: diameter of the thread root of the screw increases, thread of the screw decreases gradually, diameter of the sheath decreases, outlet of the cake is narrowed [134]. Generally, the residual liquid fraction contained in the cake issued by screw pressing cannot be recovered by pressing; a solvent extraction of the pressing cake is necessary for complete recovery.

3.6.2 Process and Procedure

Hydraulic pressing is performed in batch by a piston and achieves to reach a pressure of 500 bar on the material to be pressed. Membranes pressing could also perform batch expression where the product is placed in a perforated metallic cylinder to allow the extract to flow. The elastic membrane of cylinder is inflated by compressed air, which lead to the expression of liquid fraction contained in the product.

The continuous presses can be classified in three classes: expellers, expanders, and twin screw presses [134]. Expellers are composed of a screw, rotating inside a perforated barrel to allow flowing of the expelled oil. At the screw head, a cone partially obstructs the meal discharge, causing the necessary pressure increase to expel oil/fluids. Expander is used for pre-treatment or expression of low-oil content material (soybeans). Whereas twin screw extruders have the advantage to skip the pre-treatment phases, according to the screw configurations. However, expression by twin screw extruders has yet been limited to lab and pilot plant scale. Figure 3.14 represents different types of mechanical press for oil extraction.

3.6.3 Applications for Microbial Oil

Historically, mechanical pressing was the most common processes for recovering oil from seed oil material such as soybean or sunflower. In recent years, this process can also be used for extraction of lipids from micro-organisms such as microalgae or algae (cyanobacteria) with the aim of biofuel production [135, 136]. According to a study conducted by Topare et al. [137], expeller-pressing process can recover about 75 % of oil from various types of algae. Some preliminary step on the material are necessary to optimize this method, as well as surface algae need to be wetted with water to permit an easily movement through perforated barrel. Abbassi et al. [138] proved that oil from *Nannochloropsis oculata* biomass could be successful extracted using a manual operated hydraulic press for pressure superior to 30 bars combined with retreatment with liquid nitrogen for the disruption of cell strain. On the other hand, a simple mechanical pressing of the oleaginous biomass is

Fig. 3.14 Photography of different types of hydraulic and mechanical press

sufficient to recover lipids [139]. But Cooney et al. [140] have noted that the major drawback of mechanical pressing was the unicellular nature of microalgae cells and rigidity of some strains. In the future, improvement of pressing equipment is crucial taking in account the size of microorganisms.

3.7 Bio-Based or Agro-Solvents Extraction

3.7.1 Principle

Nowadays, the most common technique for extraction of lipids from microorganisms such as yeast, fungi, bacteria or micro-algae is generally solvent extraction, using mixtures of petroleum solvents such as chloroform/methanol or hexane. In a point of view of environmental protection and the development of green chemistry, these flammable and toxic petroleum solvents have to be replaced in the future by bio-based or so called agro-solvents. These solvents can be distinguished from "green" solvents such as dibasic esters [141] or ionic liquids [142], which are no toxic and environmental friendly, but still synthetized from petrol. Agro-solvents can be classified in three classes, according the agro-sector (Fig. 3.15): cereal/sugar, oleo-proteagineous and wood.

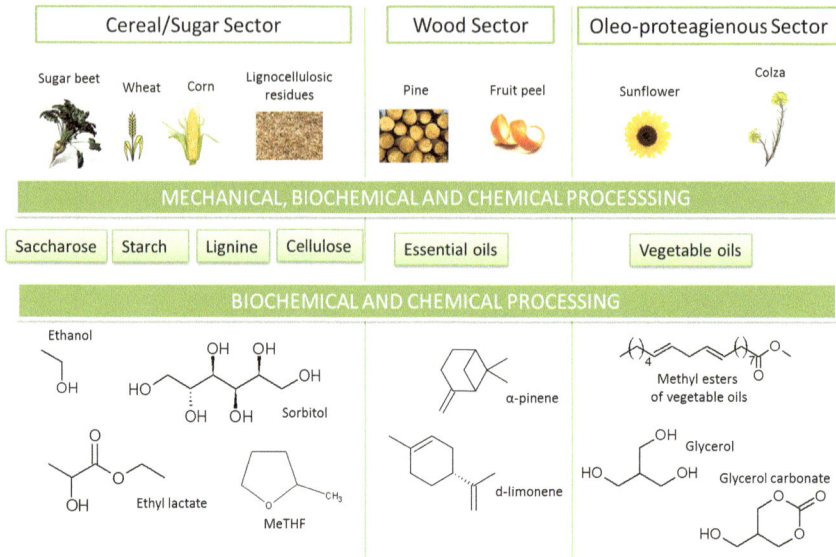

Fig. 3.15 Different classes of agro-solvent

Agro-solvents derived from cereal/sugar sector are mainly obtained from natural fermenting of sweet juices contained in plants such as sugar beet, sugarcane, wheat, corn, etc. Ethanol derived from vegetable, called bio-ethanol is produced about 60 % from sugarcane and 40 % from other crops [143].

Besides, fermentation of glucose gives many possibilities to product some molecules, which could be used as agro-solvents such as sorbitol, esters of lactic acid (ethyl lactate), derivatives of succinic acid, etc. [144]. At least, lignocellulosic residues derived from cereals production, especially from straw, can be exploited for production of furfural.

The oleo-chemistry including all oleo-proteagineous sector, allows to product solvents from seeds rich in vegetable oils, such as colza, sunflower and soya. The main agro-solvents molecules obtained are esters of fatty acids and derivatives of glycerol.

The agro-solvents obtained from wood sector are principally produced from conifers such as pines or fruit peels. The main solvents natives from this sector are terpenes hydrocarbons, $C_{10}H_{16}$ types, mono-cyclic terpene (*d*-limonène) and bi-cyclic terpene (*α*- and *β*-pinene). Limonene is most often obtained by steam-distillation of byproducts mostly peels generated by orange juice industry. *α*-pinene and *β*-pinene are produced by steam-distillation of oleoresins contained in pines or can be extract from gum turpentine, a kind of essential oil distilled from pine gum or black liquor from paper mills [145].

3.7.2 Process and Procedure

There are various effective tools, most often empirical, to classify or select solvents for a given application, according to their solvating and/or physicochemical properties. To estimate and quantify solvent power to extract a suitable compound for a given application, Hansen solubility parameters constitutes a simple empirical and qualitative guide widely used in academia and industrial research laboratories [146]. Since developed in the 1960s for the paint and varnish industries, the theory of Hansen Solubility Parameters (HSP) has been widely used in many industries: pharmaceutical, detergent, cosmetic, extraction etc. The method allows to predict the compatibility or affinity between different chemical substances and notably to estimate the capacity of agro-solvents to substitute toxic solvents derived from petrochemical [147]. To take account of other intermolecular forces, Hansen has decomposed total cohesion energy of the system in the amount of cohesive energies corresponding to the principal modes of interaction encountered in common organic materials. This decomposition allows defining three HSP parameters: δ_d, the energy from dispersion bonds between molecules, δ_p, the energy from dipolar intermolecular force between molecules and δ_h, the energy from hydrogen bonds between molecules. That therefore means that for two substances being miscible, it is necessary that their three HSP parameters are identical or very close. Representatively, HSP parameters allow access to an area of three-dimensional solubility [145].

3.7.3 Application for Microbial Oil

The first work using agro-solvents such as terpenes as an alternative to petroleum solvents for the lipid extraction from microorganisms such as micro-algae was proposed by Dejoye et al. [148]. In this study, authors have shown that lipid extraction by Soxhlet, usually performed with apolar solvent such as *n*-hexane, could be substituted by alternative solvents derived from renewable feedstock such as *d*-limonene, *α*-pinene and *para*-cymene. The extracted crude oil was higher using terpenes than *n*-hexane (0.88, 1.29, 0.91 and 1.52 g per 100 g of dry weigh biomass for *n*-hexane, *d*-limonene, *α*-pinene and *para*-cymene, respectively). This effect has already been noted by Liu and Mamidipally [148, 149] for rice bran oil extraction using *d*-limonene and hexane, this difference might be due to the slight more polar nature of terpenes compared to *n*-hexane.

Moreover, the fatty acids with chain lengths (C14–C22) composition of lipids extract from *C. vulgaris* biomass according to the terpenes and *n*-hexane solvents trends to be similar in term of quantity and quality [148], and were particularly recognized as the common fatty acids contained in biodiesel [151]. On the other hand, Dejoye [152] have developed a new procedure, called simultaneous extraction and distillation process (SDEP) using agro-solvents, for lipid extraction from wet microalgae (*N. oculata* and *Dunaliella salina*). SDEP-cymene extraction for *N. oculata* has shown a higher total lipid yield in comparison with Blight and Dyer method (21.45 and 23.78 %) and no significant differences in terms of distribution of lipids classes and fatty acid composition. More recently, Golmakani [153] has combined the use of limonene to the pressurized liquid extraction (PLE) for the extraction of lipids from various marine microorganisms. PLE at 200 °C for 15 min using limonene:ethanol (1:1, v/v) as extracting solvent to extract lipids from *Spirulina, Phormidium, Anabaena* and *Stigeoclonium* has indicated a enrichment of lipid extracts in valuable fatty acids in a short extraction time. Furthermore, agro-solvents can be used as a new alternative combined technique of biofuel production, depending on the species studied.

3.8 Enzyme-Assisted Extraction

3.8.1 Principle

Enzymes-assisted aqueous extraction processing (EAEP) have been promoted since 40 years for fractionation of biological material and for extraction of interest molecules in an economical and safe manner. Enzymes are biological tools of nature and allow accelerating metabolic reactions of a living organism. Most of enzymes are proteins, excepted in the case of ribozymes (RNA), and exhibit a biological catalysis activity and selectivity toward substrates [154]. Such as any catalyst, enzymes do not involved in the reaction process and are intact in end of the reaction. Its role is to contribute to the increasing of the kinetic of chemical reactions (10^6–10^{12} faster) without changing the equilibrium constant and decreasing the free activation energy. Enzymes which are biological catalysts differ from chemical catalysts on some points such as reactions are carried out in softer conditions in terms of pH, temperature, pressure. A greater specificity of reactions and a possible regulation of catalysis could be obtained by changing concentration in products and substrates, pH, temperature, and metabolism. There is at least one different enzyme catalyzed reaction which represents thousands of enzymes by organisms, more than 90,000 names of enzymes are listed and 27,700 of them are identified as proteic structure [155]. Generally using of enzymes (proteases and hydrolase polysaccharides) allow to enhancing extraction of lipid from microorganisms by facilitating the release of oil from the plant tissue breakdown [156]. As observed in Fig. 3.16, the solid surface has changed wetting conditions and oil is being released. A film consisting of enzyme—water phase protects the surface to become oil wet. Enzymes are used to hydrolyze structural polysaccharides of the cell wall of biological material and proteins associated with lipid bodies. Enzyme-assisted extraction (EAE) is an efficiency and recognized method employed to improve lipid extraction from several vegetable material such as soybean, sunflower, microalgae [157–159]. Enzymes can be extracted from any living organism from bacteria to fungi, plants to animals. Among all enzymes used industrially, more than half come from fungi or yeasts, about a third are bacterial and what remains is divided between animal sources or vegetable.

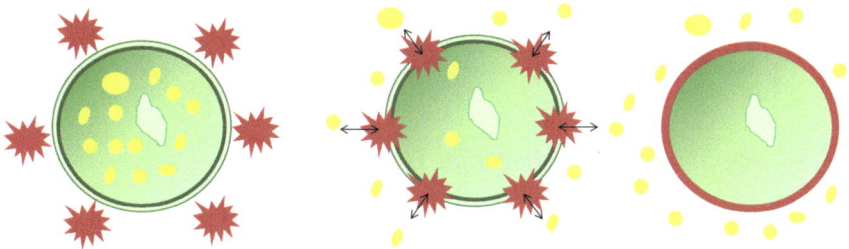

Fig. 3.16 Mechanism of enzyme-assisted aqueous extraction on biological cell

3.8.2 Process and Procedure

Proteins own various activity and types (proteases or hydrolases polysaccharides) and then become auxiliary breakdown of cell walls oil biomass tissues. The process is almost identical to the aqueous extraction (besides an enzymatic hydrolysis phase). There are no rules concerning the enzymatic activities and operation setting. This heterogeneity is linked to structural and physicochemical properties of the substrate used and the nature of the enzymes employed. Various factors can affect the efficiency of the enzymatic extraction: temperature, pH, substrate concentration and enzyme, ionic strength, the presence or absence of inhibitory substances/enhancer, and the quantity of solvent added [160].

A trend seems to be that the use of multi-enzyme complexes, which generally gives better results than a single enzyme [161, 162]. Enzymes are recently classified according to these features according to the nomenclature given by the International Union of Biochemistry and Molecular Biology (IUBMB) [163]. Historically, one enzyme could have various names and/or one name could regroup several enzymes. Nowadays, enzymes are homogenized and referenced under an E.C. (Enzyme Commission) composed by numbers. First one (x) corresponds to the enzyme class (type of chemical reaction catalyzed), the second number (y) is the subcategory (mechanism of reaction), the third one (z) corresponds to the nature of implied molecules (designate the main substrate and co-substrate of the enzyme) and the last number (t) defines the number of enzyme order (serial number) [xx]. In this classification, name of enzymes beginning by 1 defines an enzyme able to catalyze a reduction-oxidation reaction and enzymes with a number beginning by 2 can catalyze the transfer of chemical functions. The names of enzyme correspond to the substrate + reaction + suffix "ase" and are written as E.C.x.y.z.t. The generic name of enzyme is the name of reaction catalyzed by adding at the end of reaction name the suffix "ase". Major classes of enzymes are: oxidoreductases, transferases, hydrolases, lyases, isomerases and ligases. For enzyme-assisted aqueous extraction of lipid extraction from microorganisms, enzymes are generally cellulase, snailase (a complex of more than 30 enzymes, including cellulase, hemicellulase, galactase, proteolytic enzyme, pectinase, β-glucuronidase etc.), neutral protease, alkaline protease, trypsin etc. [154]. To obtain an efficient extraction by enzymes-assisted aqueous extraction processing, it is necessary to know the composition of the raw material to choice the most appropriate enzymes.

3.8.3 Application for Microbial Oil

Enzyme-assisted aqueous extraction (EAE) of microbial oil from microalgae facilitates cell disruption and lipid recovery. Taher et al. [165] showed in a study that addition of enzymes such as cellulase and trypsin to the *Scenedesmus* sp. Microalgae wet biomass may be enable enhancing the extraction of intracellular lipids after degradation of though polymers present on the cell structure. Liang et al. [164] have confirmed this phenomenon and conclude that EAE is a very efficient and rapid method for lipid extraction from various microalgae (*C. vulgaris*, *Scenedesmus dimorphus*, and *Nannochloropsis* sp.). Cellulase, neutral protease, alkaline protease are the most efficient enzymes to obtain higher lipid recovery as compared to snailase and trypsin. These authors have combined EAE with sonication treatment and find a highest lipid recovery of 49.82 %. Another study [166] have proven the interest of combination methods, microwave pre-treatment with the recombinant β-1,3 glucomannanase and ethyl acetate extraction for higher lipid recovery from oleaginous microorganisms such as *Rhodosporidium toruloides* yeast. EAE is an efficient process that could be expected to work with other oleaginous species for microbial oil application.

Chapter 4
Conclusion

The use of innovative extraction techniques such as ultrasound, microwave, instant controlled pressure drop, supercritical fluid, pulsed electric fields, mechanical pressing, agro-solvent and enzyme assisted extraction (Table 4.1) allows total recovery of lipids from microorganisms and also reduced extraction time, energy consumed, and less solvent. Conventional techniques are limited by the diffusion of solvent into biomass, due to rigid structure of cell walls of microorganisms. The solution could be to enhance the diffusion of solvent and to disrupt cell walls. For example, ultrasound and electric pulse fields allow in a high disruption of cell which permit to accelerate the mass transfer, thus recovery lipids in a short time. In another hand, heating by microwave induces combined mass and heating transfer which permits the destruction of cells and liberation of lipids.

© The Author(s) 2015   43
A. Meullemiestre et al., *Modern Techniques and Solvents for the Extraction of Microbial Oils*, SpringerBriefs in Green Chemistry for Sustainability,
DOI 10.1007/978-3-319-22717-7_4

Table 4.1 Characteristics, main disadvantages and advantages of various processes for lipids extraction from microorganisms

Name	Investment	Sample size	Extraction time	Main disadvantages	Main advantages
Maceration	Low	>1000 L	High	Limited by solubility	Large scale
Ultrasound	Low	600 L	Low	Problem for separation	High cell disruption
Microwave	Medium	150 L	Low	Hot spots	Cell disruption
DIC	High	100 L	Low	High energy consumption	High cell disruption
SFE	High	300 L	Medium	Need of know-how	Enhance mass transfer
PEF	High	Continuous	Medium	Difficult ease of operation	Electroporation of wall cells
Mechanical pressing	Low	Continuous	Low	Impact of size of microorganisms	Easy to use
Bio-based solvents	Low	>1000 L	High	Legislation	Interesting physico-chemical properties
Enzyme	Medium	>1000 L	Medium	Precise conditions	Less solvent

References

1. Madigan M, Martinko J, Dunlap P (2008) Brock biology of microorganisms. Pearson Education Inc, London
2. Boze H, Moulin G, Galzy P (1995) Production of microbial biomass. Biotechnol Set 2:165–220
3. Northcote D, Horne R (1952) The chemical composition and structure of the yeast cell wall. Biochem J 51:232
4. Blumreisinger M, Meindl D, Loos E (1983) Cell wall composition of chlorococcal algae. Phytochemistry 22:1603–1604
5. Abo-Shady AM, Mohamed YA, Lasheen T (1993) Chemical composition of the cell wall in some green algae species. Biol Plant 35:629–632
6. Wynn JP, Ratledge C (2005) Oils from microorganisms. Bailey's Industrial Oil and Fat Products
7. Spolaore P, Joannis-Cassan C, Duran E, Isambert A (2006) Commercial applications of microalgae. J Biosci Bioeng 101:87–96
8. Halász A, Lásztity R (1990) Use of yeast biomass in food production. CRC Press, Boca Raton
9. Dufosse L (2009) Pigment microbial. Encycl Microbial 4:457–471
10. Rosalam S, England R (2006) Review of xanthan gum production from unmodified starches by *Xanthomonas comprestris* sp. Enzyme Microbial Technol 39:97–207
11. Barrow C, Shahidi F (2007) Marine nutraceuticals and functional foods. CRC Press, Boca Raton
12. Lubián LM, Montero O, Moreno-Garrido I, Huertas IE, Sobrino C, González-del Valle M, Parés G (2000) *Nannochloropsis* (Eustigmatophyceae) as source of commercially valuable pigments. J Appl Phycol 12:249–255
13. Custódio L, Soares F, Pereira H, Barreira L, Vizetto-Duarte C, Rodrigues MJ, Varela J (2014) Fatty acid composition and biological activities of *Isochrysis galbana* T-ISO, *Tetraselmis* sp. and *Scenedesmus* sp: possible application in the pharmaceutical and functional food industries. J Appl Phycol 26:151–161
14. Stolz P, Obermayer B (2005) Manufacturing microalgae for skin care. Cosmet Toiletries 120:99–106
15. Venil CK, Zakaria ZA, Ahmad WA (2013) Bacterial pigments and their applications. Process Biochem 48:1065–1079
16. Malik K, Tokkas J, Goyal S (2012) Microbial pigments: a review. Int J Appl Microbiol Biotechnol Res 1:361–365
17. Hernández MS, Rodriguez MR, Guerra NP, Rosés RP (2006) Amylase production by *Aspergillus* niger in submerged cultivation on two wastes from food industries. J Food Eng 73:93–100
18. Andjelković U, Pićurić S, Vujčić Z (2010) Purification and characterisation of *Saccharomyces cerevisiae* external invertase isoforms. Food Chem 120:799–804

A. Meullemiestre et al., *Modern Techniques and Solvents for the Extraction of Microbial Oils*, SpringerBriefs in Green Chemistry for Sustainability,
DOI 10.1007/978-3-319-22717-7

19. Richardson GH, Nelson JH, Lubnow RE, Schwarberg RL (1967) Rennin-like enzyme from *Mucor pusillus* for cheese manufacture. J Dairy Sci 50:1066–1072
20. Dipak P, Lele S (2005) Carotenoid production from microalga, *Dunaliella salina*. Ind J Biotechnol 4:476–483
21. Safi C, Charton M, Pignolet O, Pontalier PY, Vaca-Garcia C (2013) Evaluation of the protein quality of *Porphyridium cruentum*. J Appl Phycol 25:497–501
22. Mendes AS, de Carvalho JE, Duarte MC, Durán N, Bruns RE (2001) Factorial design and response surface optimization of crude violacein for *Chromobacterium violaceum* production. Biotechnol Lett 23:1963–1969
23. Williams RP (1973) Biosynthesis of prodigiosin, a secondary metabolite of *Serratia marcescens*. Appl Microbiol 25:396–402
24. Wang Z, Wang Y, Xie F, Chen S, Li J, Li D, Chen X (2014) Improvement of acetoin reductase activity enhances bacitracin production by *Bacillus licheniformis*. Process Biochem 49:2039–2043
25. Noda M, Kawahara Y, Ichikawa A, Matoba Y, Matsuo H, Sugiyama M Lee DG (2004) Self-protection Mechanism in D-cycloserine-producing Streptomyces lavendulae gene cloning, characterization, and kinetics of its alanine racemase and D-alanyl-D-alanine ligase, which are target enzymes of D-cycloserine. J Biol Chem 279:46143–46152
26. Saenge C, Cheirsilp B, Suksaroge TT, Bourtoom T (2011) Potential use of oleaginous red yeast *Rhodotorula glutinis* for the bioconversion of crude glycerol from biodiesel plant to lipids and carotenoids. Process Biochem 46:210–218
27. Harder MNC, Delabio AS, Cazassa S, Remedio RR, Pires JA, Monteiro TRR, Arthur V (2013) Lipid production by *Yarrowia lipolytica* for biofuels, pp 274–278
28. Martins FF, Ferreira TF, Azevedo DA, Alice M, Coelho Z (2012) Evaluation of crude oil degradation by *Yarrowia lipolytica*. Chem Eng 27
29. Converti A, Casazza AA, Ortiz EY, Perego P, Del Borghi M (2009) Effect of temperature and nitrogen concentration on the growth and lipid content of *Nannochloropsis oculata* and *Chlorella vulgaris* for biodiesel production. Chem Eng Process 48:1146–1151
30. Doughman SD, Krupanidhi S, Sanjeevi CB (2007) Omega-3 fatty acids for nutrition and medicine: considering microalgae oil as a vegetarian source of EPA and DHA. Curr Diab Rev 3:198–203
31. Brown MR, Jeffrey SW, Volkman JK, Dunstan GA (1997) Nutritional properties of microalgae for mariculture. Aquaculture 151:315–331
32. García-González M, Moreno J, Manzano JC, Florencio FJ, Guerrero MG (2005) Production of *Dunaliella salina* biomass rich in 9-*cis*-β-carotene and lutein in a closed tubular photobioreactor. J Biotechnol 115:81–90
33. Barba N, Grimi E, Vorobiev (2014) New approaches for the use of non-conventional cell disruption technologies to extract potential food additives and nutraceuticals from microalgae. Food Eng Rev 7:45
34. Cescut J (2009) Accumulation d'acylglycérols par des espèces levuriennes à usage carburant aéronautique: physiologie et performances de procédés. Thèse de doctorat de l'Université de Toulouse
35. Ratledge C (1991) Microorganisms for lipids. Acta Biotechnol 11:429–438
36. Rattray JB, Schibeci A, Kidby DK (1975) Lipids of yeasts. Bacteriological Rev 39:197
37. Gunstone FD, Harwood JL, Dijkstra AJ (2007) The lipid handbook with CD-ROM. CRC Press, Boca Raton
38. Losel DM (1988) Fungal lipids. In: Ratledge C, Wilkinson SG (eds) Microbial lipids, vol 1. Academic Press, London, pp 699–806
39. Moss CW, Shinoda T, Samuels JW (1982) Determination of cellular fatty acid compositions of various yeasts by gas-liquid chromatography. J Clin Microbiol 16:1073–1079
40. Bellou S, Aggelis G (2013) Biochemical activities in *Chlorella* sp. and *Nannochloropsis salina* during lipid and sugar synthesis in a lab-scale open pond simulating reactor. J Biotechnol 164:318–329

41. Bellou S, Makri A (2014) Morphological and metabolic shifts of *Yarrowia lipolytica* induced by alteration of the dissolved oxygen concentration in the growth environment. Microbiology 160:807–817

42. Domergues F, Spiekermann P, Lerchil J (2003) New insight into *Phaeodactylum tricornutum* fatty acid metabolism. Cloning and functional characterization of plastidial and microsomal 12-fatty acid desaturases. Plant Physiol 131:1648–1660

43. Park JY, Oh YK, Lee JS (2014) Acid-catalyzed hot-water extraction of lipids from *Chlorella vulgaris*. Bioresour Technol 153:408–412

44. Wältermann M, Luftmann H, Baumeister D, Kalscheuer R, Steinbüchel A (2000) *Rhodococcus opacus* strain PD630 as a new source of high-value single-cell oil? Isolation and characterization of triacylglycerols and other storage lipids. Microbiology 146:1143–1149

45. Ndowora TCR, Kinkel LL, Jones RK, Anderson NA (1996) Fatty acid analysis of pathogenic and suppressive strains of *Streptomyces* species isolated in Minnesota. Phytopathology 86:138–143

46. Ishida Y, Kitagawa K, Nakayama A, Ohtani H (2006) Complementary analysis of lipids in whole bacteria cells by thermally assisted hydrolysis and methylation-GC and MALDI-MS combined with on-probe sample pretreatment. J Anal Appl Pyrol 77:116–120

47. Thevenieau F, Nicaud JM (2013) Microorganisms as sources of oils. OCL 20:D603

48. Soxhlet (1990) AOAC, Association of Official Analytical Chemist. In: Helrich K (ed) Official methods of analysis of the association of official analytical chemists, 15th edn. AOAC, Arlington

49. Bligh EG, Dyer EJ (1959) Rapid method of total lipid extraction and purification. Can J Biochem Physiol 37:911–917

50. Folch J, Lees M, Sloane Stanley GH (1957) A simple method for the isolation and purification of total lipids from animal tissues. J Biol Chem 226:497–509

51. Medina AR, Grima EM, Giménez AG, González MJI (1998) Downstream processing of algal polyunsaturated fatty acids. Biotechnol Adv 16:517–580

52. Halim R, Gladman B, Danquah MK, Webley PA (2011) Oil extraction from microalgae for biodiesel production. Bioresour Technol 102:178–185

53. Iverson SJ, Lang SLC, Cooper MH (2001) Comparison of the Bligh and Dyer and Folch methods for total lipid determination in a broad range of marine tissue. Lipids 36:1283–1287

54. Burja AM, Armenta RE, Radianingtyas H, Barrow CJ (2007) Evaluation of fatty acid extraction methods for *Thraustochytrium* sp. ONC-T18. J Agr Food Chem 55:4795–4801

55. Soxhlet F (1879) Die gewichtsanalytische Bestimmung des Milchfettes. Dingler's Polytechnisches J 232:461–465

56. Christie WW (1982) A simple procedure for rapid transmethylation of glycerolipids and cholesteryl esters. Resources 23:1072–1075

57. Morrison WR, Smith LM (1964) Preparation of fatty acid methyl esters and dimethylacetals from lipids with borontrifluoride-methanol. J Lipid Res 5:600–608

58. Dugan LR, McGinnis GW, Vadehra DV (1966) Low temperature direct methylation of lipids in biological materials. Lipids 1:305–308

59. Official Method Ce 2-66 (1989) American Oil Chemist's Society, Champaign, IL

60. Dejoye Tanzi C, Abert Vian M, Chemat F (2013) New procedure for extraction of algal lipids from wet biomass: a green clean and scalable process. Bioresour Technol 134:271–275

61. Metcalfe LD, Schmitz AA (1961) The rapid preparation of fatty acid esters from lipids for gas chromatographic analysis. Anal Chem 33:363–364

62. Christie WW (2003) Lipid analysis. The Oily Press, Bridgewater

63. Ciesla L, Waksmundzka-Hajnos M (2009) Two dimensional thin layer chromatography in the analysis of secondary plant metabolites. J Chromatogry A 1216:1035–1052

64. Momchilova S, Nikolova-Damyanova B, Waksmundzka-Hajnos M, Sherma J, Kowalska M (2008) Thin-layer chromatography in phytochemistry. CRC Press, Boca Raton

65. Fuchs B, Süss R, Teuber K, Eibisch M, Schiller J (2011) Lipid analysis by thin-layer chromatography-a review of the current state. J Chromatogry A 1218:2754–2774

66. Mason TJ (1990) Chemistry with ultrasound. Elsevier Applied Science, London
67. Pingret D, Fabiano-Tixier AS, Chemat F (2012) Accelerated methods for sample preparation in food. Comprehensive Sampling and Sample Preparation, pp 441–455
68. Mason TJ, Lorimer JP (2002) Applied sonochemistry. Wiley, Weinhein
69. Adam F, Nikitenko S, Chemat F (2011) Extraction assistée par ultrasons. Eco-extraction du végétal. Dunod, Paris, pp 91–117
70. Lagha A, Chemat S, Bartels P, Chemat F (1999) Microwave—ultrasound combined reactor suitable for atmospheric sample preparation procedure of biological and chemical products. Analysis 27:452–457
71. Hu A, Zhao S, Liang H, Qiu T, Chen G (2007) Ultrasound assisted supercritical fluid extraction of oil and coixenolide from adlay seed. Ultrason Sonochem 14:219–224
72. Luque-García JL, Luque de Castro MD (2004) Ultrasound-assisted Soxhlet extraction: an expeditive approach for solid sample treatment. J Chromatogr A 1034:237–242
73. Chemat F, Huma Z, Khan MK (2011) Applications of ultrasound in food technology: processing, preservation and extraction. Ultrason Sonochem 18:813–835
74. Adam F, Abert-Vian M, Peltier G, Chemat F (2012) Solvent-free ultrasound-assisted extraction of lipids from fresh microalgae cells: a green, clean and scalable process. Bioresour Technol 114:457–465
75. Wang C, Chen L, Rakesh B, Qin Y (2012) LVR Front Energy 6(3):266–274
76. Zhang X, Yan S, Tyagi RDT, Drogui P, Surampalli RY (2014) Ultrasonication assisted lipid extraction from oleaginous microorganisms. Bioresour Technol 158:253–261
77. Prabakaran P, Ravindran AD (2011) A comparative study on effective cell disruption methods for lipid extraction from microalgae. Lett Appl Microbiol 53(2):150–164
78. Araujo GS, Matos LJBL, Fernandes JO, Cartaxo SJM, Goncalves LRB, Fernandes FAN, Farias WRL (2013) Extraction of lipids from microalgae by ultrasound application: prospection of the optimal extraction method. Ultrason Sonochem 20:95–98
79. Bermudez Menendez JM, Arenillas A, Menedez Diaz JA, Boffa L, Mantegna S, Binello A, Cravatto G (2014) Optimization of microalgae oil extraction under ultrasound and microwave irradiation. J Chem Technol Biotechnol 89:1779–1784
80. Halim R, Rupasinghe TWT, Tull DL, Webley PA (2013) Mechanical cell disruption for lipid extraction from microalgal biomass. Bioresour Technol 140:53–63
81. Gerde JA, Montalbo-Lomboy M, Yao L (2012) Evaluation of microalgae cell disruption by ultrasonic treatment. Bioresour Technol 125:175–181
82. Ganzler K, Salgó A, Valkó K (1986) Microwave extraction. A novel sample preparation method for chromatography. J Chromatogr A 371:299–306
83. Lane D, Jenkins SWD (1984) Presented at the 9th international symposium on polynuclear aromatic hydrocarbons, Columbus, OH, Abstracts 437
84. Pingret D, Fabiano-Tixier AS, Chemat F (2012) Accelerated methods for sample preparation in food. Comprehensive sampling and sample preparation. Academic Press, Oxford, pp 441–455
85. Dejoye Tanzi C, Abert Vian M, Chemat F (2013) New procedure for extraction of algal lipid from wet biomass: a green clean and scalable process. Bioresour Technol 134:271–275
86. Bermudez Menendez JM, Arenillas A, Menedez Diaz JA, Boffa L, Mantegna S, Binello A, Cravatto G, JM (2014) Optimization of microalgae oil extraction under ultrasound and microwave irradiation. J Chem Technol Biotechnol 89:1779–1784
87. Paré JRJ, Sigouin M, Lapointe J (1990) Extraction de produits naturels assistés par micro-ondes. Eur Patent EP 398798
88. Luque de Castro MD, Priego-Capote F (2011) Microwave-assisted extraction enhancing extraction processes in the food industry. CRC Press, Cambridge, pp 85–122
89. Luque-García JL, Luque de Castro MD (2004) Focused microwave-assisted Soxhlet extraction: devices and applications. Talanta 64:571–577

90. Virot M, Tomao V, Colnagui G, Visinoni F, Chemat F (2007) New microwave-integrated Soxhlet extraction: an advantageous tool for the extraction of lipids from food. J Chromatogr A 1174:138–144

91. Sparr Eskilsson C, Björklund E (2000) Analytical-scale microwave-assisted extraction. J Chromatogr A 902:227–250

92. Chuck CJ, Lou-Hing D, Dean R, Sargeant LA, Scott RJ, Jenkins RW (2014) Simultaneous microwave extraction and synthesis of fatty acid methyl ester from the oleaginous yeast *Rhodotorula glutinis*. Energy 69:446–454

93. Khoomrunng S, Chumnanpuen P, Jansa-Ard S, Stahlman M, Nookaew I, Boren J, Nielsen J (2013) Rapid quantification of yeast lipid using microwave-assisted total lipid extraction and HPLC-CAD. Anal Chem 85:4912–4919

94. Prabakaran P, Ravindran AD (2011) A comparative study on effective cell disruption methods for lipid extraction from microalgae. Lett Appl Microbiol 53(2):150–164

95. Ryckebosch E, Bermúdez SPC, Termote-Verhalle R, Bruneel C, Muylaert K, Parra-Saldivar R, Foubert I (2014) Influence of extraction solvent system on the extractability of lipid components from the biomass of *Nannochloropsis gaditana*. J Appl Phycol 26(3):1501–1510. doi:10.1007/s10811-013-0189-y

96. Allaf K, Vidal P (1989) Feasibility Study of a Process of Drying/Swelling by Instantaneous Decompression Toward Vacuum of in Pieces Vegetables in View of Rapid Re-Hydration Gradient Activity Plotting. University of Technology of Compiègne UTC N°CR/89/103 industrial SILVALAON Partner

97. Allaf K, Rezzoug SA, Cioffi F, Contento MP (1998) Fr Patent 98/11105

98. Allaf K, Besombes C, Berka B, Kristiawan M, Sobolik V, Allaf T (2011) Enhancing extraction processes in the food industry. CRC Press, Boca Raton, pp 255–302

99. Haddad J, Louka N, Gadouleau M, Juhel F, Allaf K (2001) Application du nouveau procédé de séchage/texturation par Détente Instantanée Contrôlée (DIC) aux poisson: impact sur les caractéristiques physicochimique du produit fini (New process of drying/texturizing by Controlled Instantaneous Pressure Drop (DIC). Application on fish: impact on the physicochemical characteristics of the final product). Sciences des aliments 21:481–498

100. Maache-Rezzoug Z, Maugard T, Nouviaire A, Pierre G, Rezzoug SA (2011) Optimizing thermomechanical pretreatment conditions to enhance enzymatic hydrolysis of wheat straw by response surface methodology. Biomass Bioenergy 35:3129–3138

101. Nguyen Van C (2010) Maîtrise de l'aptitude technologique des oléagineux par modification structurelle : applications aux opérations d'extraction et de transestérification in-situ. Thèse de l'Université de La Rochelle

102. Allaf T, Fine F, Tomao V, Nguyen C, Ginies C, Chemat F (2014) Impact of instant controlled pressure drop pre-treatment on solvent extraction of edible oil from rapeseed seeds. OCL 21(3):A301

103. Kamal I, Besombes C, Allaf K (2014) One-step processes for in-situ transesterification to biodiesel and lutein extraction from microalgae *Phaeodactylum* using instant controlled pressure drop (DIC). In: GPE—4th international congress on green process engineering, Sevilla

104. Brunner G (2005) Supercritical fluids: technology and application to food processing. J Food Eng 67:21–33

105. Cagniard de la Tour Ch (1823) Annales de Chimie Physique MM Gay-Lussac & Arago. Librairie Crochard, Paris 23:267–269

106. Hecht E (1996) Physics Calculus, Brooks/Cole. A division of international Thomson publishing Inc 14:548

107. Hannay JB, Hogarth J (1880) On the solubility of solids in gases. Proc R Soc Lond 30: 178–188

108. Mukhopadhyay M (2009) Extraction and processing with supercritical fluids. J Chem Technol Biotechnol 84:6–12

109. Rozzi NL, Singh RK (2002) Supercritical fluids and the food industry. Compr Rev Food Sci Food Saf 1:33–44
110. Milanesio J, Hegel P, Medina-Gonzalez Y, Camy S, Condoret JS (2013) Extraction of lipids from *Yarrowia Lipolytica*. J Chem Technol Biotechnol 88:378–387
111. Hegel PE, Camy S, Destrac P, Condoret JS (2011) Influence of pretreatments for extraction of lipids from yeast by using supercritical carbon dioxide and ethanol as cosolvent. J Supercrit Fluids 58:68–78
112. Tang S, Qin C, Wang H, Li S, Tian S (2011) Study on supercritical extraction of lipids and enrichment of DHA from oil-rich microalgae. J Supercrit Fluids 57:44–49
113. Nobre BP, Villabolos F, Barragan BE, Oliveira AC, Batista AP, Marques PASS, Mendes RL, Sovova H, Palavra AF, Gouveia L (2013) A biorefinery from *Nannochloropsis* sp. *Microalga* —Extraction of oils and pigments. Production of biohydrogen from the leftover biomass. Bioresour Technol 135:128–136
114. Taher H, Al-Zuhair S, Al-Marzouqi AH, Haik Y, Farid M (2014) Effective extraction of microalgae lipids from wet biomass for biodiesel production. Biomass Bioenergy 66:159–167
115. Mendes R (2008) Supercritical fluid extraction of nutraceuticals and bioactive compounds, ed Martinez JL. CRC Press, Boca Raton, FL
116. Jeyamkondan S, Jayas DS, Holley RA (1999) Pulsed electric field processing of foods: a review. J Food Protect 62:1088–1096
117. Doevenspeck H (1961) Influencing cells and cell walls by electrostatic impulses. Fleischwirtschaft 13:968–987
118. Sale A, Hamilton W (1967) Effect of high electric fields on microorganism. Killing of bacteria and yeast. Biochim Biophys Acta 148:781–788
119. Weaver JC, Chizmadzehev YA (1996) Theory of electroporation: a review. Bioelectrochem Bioenerg 41:135–160
120. Pakhomov AG, Miklavcic D, Markov MS (eds) Advanced electroporation techniques in biology and medicine. CRC Press, Boca Raton
121. Lebovka NI, Bazhal MI, Vorobiev EI (2001) Pulsed electric field breakage of cellular tissues: visualization of percolative properties. Innov Food Sci Emerg 2:113–125
122. Lebovka L, Vorobiev E, Chemat F (2011) Enhancing extraction processes in the food industry. CRC Press, Boca Raton, pp 25–84
123. Miklavcic D, Towhidi L (2010) Numerical study of the electroporation pulse shape effect on molecular uptake of biological cells. Radiol Oncol 44:34–41
124. Barbosa-Canovas GV, Altunakar B (2006) Pulsed electric fields processing of foods: an overview. In: Raso J, Heinz V (eds) Pulsed electric field technology for food industry: fundamentals and applications. Springer, New York, pp 153–194
125. Wouters PC, Smelt JPPM (1997) Inactivation of microorganisms with pulsed electric fields: potential for food preservation. Food Biotechnol 11:193–229
126. Foltz G (2012) Algae lysis with pulsed electric fields. California Polytechnic State University, San Luis Obispo
127. Zbiden MDA, Sturm BSM, Nord RD, Carey WJ, Moore D, Shinogle H, Stagg-Williams SM (2013) Pulsed electric field (PEF) as an intensification pretreatment for Greener Solvent Lipid extraction from microalgae. Biotechnol Bioeng 110:1605–1615
128. Goettel M, Eing C, Gusbeth C, Straessner R, Frey W (2013) Pulsed electric field assisted extraction of intracellular valuable microalgae. Algal Res 2:401–408
129. Sheng J, Vannela R, Rittmann BE (2011) Evaluation of cell-disruption effects of pulsed-electric-field treatment of synechocystis PCC 6803. Environ Sci Technol 45 (8):3795–3802
130. Schwartzberg HG (1997) Expression of fluid from biological solids. Sep Purif Rev 26:1–213
131. Demirbas A (2009) Production of biodiesel from algae oils. Energy Source Part A: Recovery Utilization Environ Effects 31:163–168

132. Ramesh D (2013) Lipid identification and extraction techniques. In: Bux F (ed) Biotechnological applications of microalgae: biodisel and value-added products, CRC Press, Boca Raton, FL, pp 89–97
133. Lanoisellé JL, Bouvier JM (1994) Le pressage hydraulique des oléagineux. Mise au point. Revue Française Corps gras 41:61–74
134. Savoire R, Lanoisellé JL, Vorobiev E (2012) Mechanical continuous oil expression from oilseeds: a review. Food Bioprocess Technol 6:1–16
135. Richmond A (2004) Principles for attaining maximal microalgal productivity in photobioreactors: an overview. Hydrobiologia 512:33–37
136. Rawat I, Kumar R, Mutanda T, Bux F (2013) Biodiesel from microalgae: a critical evaluation from laboratory to large scale production. Appl Energy 103:444–467
137. Topare NS, Raut SJ, Renge VC, Khedar SV, Chavan YP, Bhagat SL (2011) Extraction of oil from algae by solvent extraction and oil expeller method. Int J Chem Sci 9(4):1746–1750
138. Abbassi A, Ali M, Watson IA (2014) Temperature dependency of cell wall destruction of microalgae with liquid nitrogen pretreatment and hydraulic pressing. Algal Res 5:190–194
139. Kleinschmidt P (2010) Methods and micro economy of biodisel production—new. Example thourgh a business plan analysis for a biodiesel plant. Slovak University of Technology
140. Cooney MJ, Young G, Pate R (2011) Bio-oil from photosynthetic microalgae: case study. Bioresour Technol 102:166–177
141. Massawe E, Geiser K, Ellenbecker M (2008) Technical performance evaluation of the potential biobased floor strippers. J Cleaner Prod 16:12–21
142. Rogers RD, Seddon KR (2003) Ionics liquids as green solvents: progress and prospects, ACS Symposium Series 856
143. Demirbas A (2005) Bioethanol from cellulosic materials: a renewable motor fuel from biomass. Energy Sour 27:327–333
144. Fine F, Abert Vian M, Fabiano Tixier AS, Carré P, Pages X, Chemat F (2013) Les agro-solvants pour l'extraction des huiles végétales issues de graines oléagineuses. Oilseeds Fats Crops Lipids 20(5):A502
145. Meier GJ (1993) Non-aqueous cleaning solvent substitution. In: National technology transfer conference and exposition (4th), United States Department of Energy, Washington, pp 223–231
146. Hansen CM (1967) The three dimensional solubility parameter—Key to paint component affinities II. Dyes, emulsifiers, mutual solubility and compatibility, and pigments. J Paint Technol 39:505–510
147. Mouloungui Z, Pelet S, Eychenne V (2006) Lipochimie. In La chimie verte. Paris: Lavoisier, Tec&Doc
148. Dejoye Tanzi C, Abert Vian M et al (2012) Terpenes as green solvents for extraction of oil from microalgae. Molecules 17:8196–8205
149. Liu SX, Mamidipally PK (2005) Quality comparison of rice bran oil extracted with d-limonene and hexane. Cereal Chem J 82:209–215
150. Mamidipally PK, Liu SX (2004) First approach on rice bran oil extraction using limonene. Eur J Lipid Sci Technol 106:122–125
151. Demirbas A (2009) Progress and recent trends in biodiesel fuels. Energy Convers Manage 50:14–34
152. Dejoye Tanzi C (2013) Eco-Extraction et analyse des lipides de micro-algues pour la production d'algo-carburant. Thèse de doctorat de l'Université d'Avignon et des Pays de Vaucluse
153. Golmakani MT, Mendiola JA, Rezaei K, Ibanez E (2014) Pressurized limonene as an alternative bio-solvent for the extraction of lipids from marine microorganisms. J Supercrit Fluids 92:1–7
154. Parkin KL (2009) General characteristics of enzymes. In: Nagodawithana T, Reed G (eds) Enzymes in food processing. Academic Press, New York, pp 7–37

155. Jung S, Nobrega de Moura JML, Campbell KA, Johnson LA (2011) Enzyme-assisted extraction of oilseeds. Enhancing extraction processes in the food industry. CRC Press, Cambridge, pp 477–518
156. Choa HS, Oha YK, Parka SC, Leeb JW, Parka JY (2013) Effects of enzymatic hydrolysis on lipid extraction from *Chlorella vulgaris*. Renew Energy 54:156–160
157. Shankar D, Agrawal YC, Sarkar BC, Singh BPN (1997) Enzymatic hydrolysis in conjunction with conventional pretreatments to soybean for enhanced oil availability and recovery. J Am Oil Chemists' Soc 74:1543–1574
158. Sineiro J, Dominguez H, Nunez MJ, Lema JM (2011) Optimization of the enzymatic treatment during aqueous oil extraction from sunflower seeds. Food Chem 61:467–474
159. Gerken HG, Donohoe B, Knoshaug EP (2013) Enzymatic cell wall degradation of *Chlorella vulgaris* and other microalgae for biofuels production. Planta 237(1):239–253
160. Dumay J (2006) Extraction de lipides en voie aqueuse par bioréacteur enzymatique combiné à l'ultrafiltration: Application à la valorisation de coproduits de poisson (Sardina pilchardus). Thèse de doctorat de l'Ecole Polytechnique de l'Université de Nantes
161. Olsen HS (1994) Aqueous Enzymatic Extraction of Oil from Rapeseeds. Manufacture of food and beverage. http://www.p2pays.org/ref/10/09365.htm
162. Barrios VA, Olmos DA, Noyola RA, Lopez-Munguia CA (1990) Optimization of an enzymatic process for coconut oil extraction. Oléagineux 45:35–42
163. Webb EC (1992) Enzyme Nomenclature. Recommendations of the Nomenclature Committee of the International Union of Biochemistry and Molecular Biology on the Nomenclature and Classification of Enzymes NC-IUBMB. Academic Press, New York
164. Liang K, Zhang Q, Cong W (2012) Enzyme-Assisted aqueous extraction of lipid from microalgae. J Agric Food Chem 60:11771–11776
165. Taher H, Al-Zuhair S, Al-Marzouqi AH, Haik Y, Farid M (2014) Effective extraction of microalgae lipids from wet biomass for biodiesel production. Biomass Energ 66:159–167
166. Jin G, Yang F, Hu C, Shen H, Zhao ZK (2012) Enzyme-assisted extraction of lipids directly from the culture of the oleaginous yeast *Rhodosporidium toruloides*. Bioresour Technol 111:378–382